# Sustainable Urban Futures

**Series Editors**
Zaheer Allam
Le Hochet
Morcellement Raffray
Terre-Rouge, Mauritius

Sina Shahab
School of Geography and Planning
Cardiff University
Cardiff, UK

This series includes a broad range of Pivot length books offering accessible and applied texts designed to appeal to both practitioners and academics in the field. Pivots in the series will explore how sustainability can be achieved in Future Cities and how technology can assist in supporting sustainable transitions to better respond to the urgencies of climate change, equity needs and inclusivity aligning the two core themes of Urban Science and Future Science.

Phillip B. Roös

# A Biophilic Pattern Language for Cities

Creating Healthy Urban Environments

Phillip B. Roös
School of Architecture and Built Env
Deakin University
Geelong, VIC, Australia

ISSN 2730-6607                     ISSN 2730-6615   (electronic)
Sustainable Urban Futures
ISBN 978-3-031-19070-4        ISBN 978-3-031-19071-1   (eBook)
https://doi.org/10.1007/978-3-031-19071-1

This Palgrave Macmillan imprint is published by the registered company Springer Nature Switzerland AG.
The registered company address is: Gewerbestrasse 11, 6330 Cham, Switzerland

*To my dearest wife Pam, and my*
*children and grandchildren*
*To the late Christopher Alexander—the father of Pattern Languages*
*and to all those other great minds who continue his work*

# ACKNOWLEDGEMENT OF COUNTRY

I want to acknowledge and pay my respect to the Elders, past, present and future, of the *Wadawurrung* people who are the Traditional Owners of the *Country* upon which this book and substance originates. Further, I want to acknowledge and pay my respect to the Traditional Owners of the lands and waters across Australia, and the Torres Strait Islander peoples and the rich cultural and intrinsic connection they have to their *Country*. I also recognise and acknowledge the contribution that First Nation Peoples have made and will continue to make extraordinary contributions to all aspects of life including culture, economy, arts and science.

# CONTENTS

# ABOUT THE AUTHOR

**Phillip B. Roös** is an academic and architect, designer, artist, writer and philosopher. His work spans a transdisciplinary discourse in the convergence of design, science, art, philosophy and environmental research. Deeply embedded in ecological consciousness, he investigates our innate affiliation to Nature—biophilia—through analysing the phenomena of living structures. His practice is positioned at the intersections of rigorous academic scholarship and applied real-world projects. As a transdisciplinary practitioner, he investigates questions of human consciousness as well as global social and environmental issues. He currently holds the position of Associate Professor in Architecture—Ecological Design at the School of Architecture and Built Environment and is the Director of the Deakin Biophilia Lab, at Deakin University. He recently authored: *Regenerative-Adaptive Design for Sustainable Development—A Pattern Language Approach* (2021). Sustainable Development Goals Series, Springer International, DOI: https://doi.org/10.1007/978-3-030-53234-5_1.

# LIST OF FIGURES

# LIST OF TABLES

CHAPTER 1

# A Biophilic Pattern Language for Cities

**Abstract** In the wake of a rapid urbanisation at a global scale, the question we need to ask ourselves is what we can do to deal with issues and problems resulting from this global trend. It is estimated that by 2050 more than 75% of the world's global population will reside in cities. There are many problems arising from this trend of urbanisation, but for the purposes of the narrative of this book, there are two issues to be dealt with: the decline of human health and wellbeing, and the decline of planetary wellbeing. *A Biophilic Pattern Language for Cities* provides guidance for the design and planning of sustainable cities by using a pattern language method embedded in the principles of biophilia. This introductory chapter charters how to use the *meta biophilic patterns* in relation to other principles in creating healthy urban environments.

**Keywords** Healthy cities • Sustainable cities • Pattern language • Biophilic design • Healthy urban environments • Biophilia

**Fig. 1.1** Biophilic patterns—Barwon Heads. (Image author: PB Roös, 2022)

## Introduction

One of our biggest global issues is rapid urbanisation and the decline of rural areas (Fig. 1.1). It is estimated that by **2050** more than 75% of the world's global population will reside in cities. The world population growth is estimated to increase at a rapid rate, reaching an estimated incredible 9.7 billion people by 2050 (UN, 2019). An increase in population results in an increase of resource needs, and according to the UN Environment Programme (UNEP) panel, the world is already running out of cheap and essential non-renewable resources such as oil, copper and gold, which in turn require rising volumes of fuel and potable water to produce (PhysOrg, 2011). The accumulated environmental impacts on Earth from rapid urbanisation is a major concern, and the resources of the planet are depleted at a rapid rate to feed the billions of humans living in cities. According to Girardet (2015) as recently as in 2012 the world's global civilisation used more natural resources in eight months than the Earth can produce, and according to current trends we will need more than two Earths to supply the needy human society with biological resources by 2030 (Girardet, 2015).

The 'writing is on the wall', and we can clearly see that the trend of global urbanisation is a phenomenon—now and will be in the future. There are many problems arising from this trend of urbanisation, but for the purposes of the narrative of this book there are two issues to highlight as a result of rapid urbanisation and city living:

1. Destruction of wild nature, unprecedented biological resource use, the extinction of species of fauna and flora at a rapid rate, the loss of biodiversity and a disconnect between humans and nature.
2. Decline of human health and wellbeing, and the decline of planetary wellbeing.

Clearly we cannot stop rapid urbanisation, the question we then need to ask ourselves is what we can do to deal with issues and problems resulting from this global trend.

The purpose of this book is to put forward a proposition to deal with one issue as noted above—the decline of human health and wellbeing, and the decline of planetary wellbeing. Acknowledging that this is a huge and complex topic to deal with, and that there are many elements to be investigated, in this instance we will look at the issues at a city level, city design, urban habitation and potential design processes to reconnect city dwellers with nature. In this book I propose a design system to be used based on the innate human-nature affiliation known as *biophilia*. A proposition has been put forward that biophilic design (using a pattern language method), will support sustainable development, in fact—embracing biophilia—will help us to progress beyond sustainability and be able to move towards a regenerative-adaptive future (Roös, 2021). This method is the *Biophilic Pattern Language for Cities* aiming to guide designers, planners and city makers to create healthy urban environments.

## SUSTAINABLE DEVELOPMENT AND BIOPHILIC DESIGN

Sustainability is important. However, considering the vast reaches across the Earth's terrain, the ecological footprint impacts of the world's global population of approximately 7.6 billion in 2018 (Roser et al., 2019 [2013]) and rapidly growing, there is an urgent need to address this issue both at a local and global scale. The most accepted and used framework to address the issues we face as humanity, is the United Nations—*Sustainable Development Goals*—mostly referred to as the SDGs (UN, 2015). In Chap. 2, I further progress the discussion of how we need to move beyond sustainability and include the 'integral sustainable' argument in this matter. Here it is worthwhile discussing the relevance of the SDGs, more specifically the SDGs that have a direct relationship to the topic of this book— SDG 3 (Good Health and Wellbeing), and SDG 11 (Sustainable Cities and Communities).

By now we know that the environments we live in work and engage with the communities around us and have a direct impact on our health and wellbeing. These places are our places of human habitation, and they need to be sustainable environments that are good for our physical and mental health. The inclusion of nature into our built urban environments creates places that can assist with human wellbeing (Fig. 1.2). However, in literature and in practice these two contexts of 'sustainability' and 'health and wellbeing' are dealt with separately, as if in silos and totally disconnected from one another. I acknowledge attempts to look at these collectively, but these attempts are minimal in the scheme of things.

Whereas the SDG 3 (Good Health and Wellbeing) related targets and indicators mostly discuss 'health and wellbeing' in relation to threatening diseases, global health, various aspects of a healthy life and health services—one of the 'related SDGs' referred to is SDG 11 (Sustainable Cities and Communities). The UN acknowledges that there is a direct interconnection between sustainable development and health, and noted in the United Nations on Environment and Development—Agenda 21 (1992), that "health and development are intimately interconnected, and called for action items under Agenda 21 be addressed according to the primary health needs of the world's population since they are integral to the achievement of the goals of sustainable development and primary environmental care" (UN, 1992, p. 31). In the contexts of achieving 'sustainable

**Fig. 1.2** Abundance of wild nature. (Image author: PB Roös, 2022)

development' and 'primary environmental healthcare', actions to realise these can be framed as the design and planning of our cities that take the environment (nature) into consideration and enhancing human and planetary health and wellbeing through acknowledging the benefits of *biophilia*, by achieving sustainable outcomes through the application of *biophilic design*.

## A Pattern Language Methodology

The structure of the four chapters in this book that deal with *four meta biophilic patterns* has been developed to follow a similar methodology to *A Pattern Language* as defined by Christopher Alexander (Alexander et al., 1977; Alexander, 1979). A typical pattern language is a method of describing a set of patterns or good design and planning processes with useful association in a specific field of expertise (Roös, 2021). In this context the *Biophilic Pattern Language for Cities* provides guidance and a framework by describing a set of patterns and design and planning processes for city making in a useful, organised and pragmatic way.

Christopher Alexander is known as the 'father of pattern language' and unquestionably the primary properties of a pattern language were derived from Alexander's (1979) argument, whereby he questioned the creation of living structures and the need for a code for the acts of design and building:

> So, I began to wonder if there was a code, like the genetic code, for human acts of building? Is there a fluid code, which generates the 'quality without a name' in buildings, and make things live? Is there some process which takes place inside a person's mind, when he allows himself to generate a building or a place which is alive? And is there indeed a process which is so simple too, that all the people of society can use it, and so generate not only individual buildings, but whole neighbourhoods and towns? It turns out that there is. It takes the form of a language (Alexander, 1979, p. 166).

The core of the pattern language system is to solve problems that occur repeatedly in the environment. To deal with this complexity of reoccurrence it is necessary to use a simple, straightforward structure to represent the pattern, and the combination of patterns. The writings of Alexander's patterns (Alexander et al., 1977) consists of a specific order, written in a specific style:

1. A *pattern title*, written in CAPITALS (Small Caps).
2. Main *representative image*, representing the nature of the configuration and adding a qualitative dimension to the problem or matter at hand.
3. *Upward links*, connecting the pattern to 'upper' patterns that may contain it and others.

4. *Problem statement* in bold, giving the configuration of the problem.
5. *Discussion*, assessing the associated issues to be discussed and addressed.
6. *Solution statement* in bold, describing the configuration of the solution.
7. *Diagram*, a sketch or cartoon of the solution with simple notes.
8. *Downward links*, connecting the pattern to 'lower' patterns that it may contain.

This system allows for embedded relationships in a hierarchical order, with overlaps, cross-linkages and uncertainties. As the patterns combine into a system, into a 'language', these overlapping relationships create a stronger, refined and more complex kind of structure that is evident in the systems of nature (Roös, 2021).

Therefore to keep in the spirit of pattern language writing, the structure of the four chapters in this book that deal with *four meta biophilic patterns* follows a similar structure to that of *A Pattern Language* (Alexander et al., 1977), and is structured as follows:

1. *Title* of the Chapter, including the *Pattern Name and Number*.
2. Main *Representative Image*, which conveys the overall message from the chapter in a visual form.
3. *Introduction*, that provides background information and a synopsis.
4. *Pattern Statement*, which sets out the context and issue to be addressed.
5. *Discussion*, the main body text that provides details of the topic and pattern.
6. *Biophilic Attributes*, that describes the supported attributes for the pattern.
7. *Pattern Applications*, note the possible options for application of the pattern.
8. *Pattern Diagram*, a sketch with simple notes about the pattern.
9. *Pattern Links*, list interconnections to other patterns and associations.
10. *Conclusion Statement*, providing a summary and statement of the pattern.

As words with grammatic relationships to one another form a spoken language, so do patterns relate to each other in a pattern language. In a pattern language the patterns are part of a directed acyclic graph, each mode of which represents a pattern. Core to the workings of a pattern language is the hierarchy of these patterns (Borchers, 2008; in Neis et al., 2012, p. 92), linked to each other in levels of scale. This is demonstrated in *A Pattern Language* (1977), where the language has a hierarchal structure within a network. This network has a sequence and going through the collection of patterns the sequence always moves from:

> ... the larger patterns to the smaller, always from the ones which create structures, to those ones which then embellish those structures, and then to those which embellish the embellishments... (Alexander et al., 1977, p. xviii).

When using *A Biophilic Pattern Language for Cities*, the *meta biophilic patterns*, their supported *biophilic attributes* and listed *related patterns from other sources*—these need to be read in a manner of understanding and using pattern languages. In essence, it is a network structure with upward links and downward links similar to what occurs in the structures of language.

## USE OF THE BIOPHILIC PATTERN LANGUAGE FOR CITIES

As described above, the *Biophilic Pattern language for Cities* has a network structure. In essence everything is connected, linked in levels of scale. The patterns in this book are 'meta' patterns with supportive biophilic attributes to support the implementation of the biophilic patterns in city planning and design. However, as per the recommendations in *A Pattern Language* (Alexander et al., 1977), the reader is encouraged to use these *four meta biophilic patterns* to develop their own patterns at different levels and scales. You will note that each meta biophilic pattern is linked to other patterns and entities within the structure of this book, as well as with other patterns and criteria in biophilic design in other literature. Each pattern has a three-part rule, which expresses a relationship between certain contexts, a problem or issue and a solution embedded in both the pattern and its entities (Alexander, 1979, p. 247). Since the language is in fact a *webstring[1] network*, there is no one sequence that perfectly captures it (Roös, 2021). This provides the opportunity for the designer or planner to dynamically develop solutions for design and planning challenges in the urban environment.

This dynamic process is evident in a hierarchical flow of the network. The sequence in the network can flow both ways between the *upward* and the *downward links*, and interconnect between the other entities such as the *biophilic attributes*. Working with the network links external to this book, connections can be created between the *meta biophilic patterns* and other biophilic patterns listed in other literature.

The reader will notice simple sketches and illustrations in the chapters of the *four meta biophilic patterns*. Their purpose is to visually represent the pattern, attributes or related entities in a simple form while attempting to address the gaps in the traditional language of words and writing, which is inadequate to describe the complexities of the phenomena behind the application of a pattern language. As Alexander (1964) puts it:

> The idea of a diagram [of a pattern] is very simple. It is an abstract pattern of physical relationships, which resolve a small system of interacting and conflicting forces, and is independent of all other forces, and of all other possible diagrams. The idea that is possible to create such abstract relationships one at a time, and to create designs, which are whole by fusing these relationships …. (Alexander, 1964, p. i)

In the spirt of design thinking, when developing opportunities to apply the *meta biophilic patterns* in projects, the reader is encouraged to embark on a journey that includes using the tool that we as humans have used since ancient times, that is the ability to create shapes and forms through the practice of drawing. We have the ability to project our ideas and thoughts onto a surface by drawing, this ability is embedded in our evolution as humans. The earliest drawing known was created by the 'modern

---

[1] Michael J. Cohen (2008) developed a model that he named the *Webstring Natural Attraction Model*, where a *webstring network* in essence demonstrates the whole interconnected strings of attachment between all entities in Nature, including humans (Cohen, 2008, p. 12). When one string in this web of life is cut, it influences and has consequences for all other entities connected in this network. What is interesting about this *webstring network* is that there is no specific sequence of the connections, but they follow a line and form their own sequence upwards and downwards in the network. This principle of upward links and downward links corresponds to the principles of *A Pattern Language* of Christopher Alexander (1977).

human' dated some 73,000 years ago (Henshilwood et al., 2018). Connecting back to nature through drawing will help us to go to first principles to link to the biophilic patterns of place, and as Alexander stated above—"*to create designs, which are whole by fusing these relationships*".

## FOUR META BIOPHILIC PATTERNS DIAGRAM (FIG. 1.3)

DIRECT EXPERIENCE OF NATURE

INDIRECT EXPERIENCE OF NATURE

EXPERIENCE OF PLACE, SPACE AND ATTACHMENT

NATURE PATTERNS, PROCESSES AND SYSTEMS

**Fig. 1.3** Sketch of the four meta biophilic patterns. (Image author: PB Roös, 2022)

## REFERENCES

Alexander, C. (1964). *Notes on the synthesis of form.* Cambridge, MA: Harvard University Press.

Alexander, C. (1979). *The timeless way of building.* Oxford University Press.

Alexander, C., Ishikawa, S., Silverstein, M., Jacobson, M., Fiksdahl-King, I., & Angel, S. (1977). *A pattern language: Towns, buildings, construction.* Oxford University Press.

Borchers, J. (2008). A pattern approach to interaction design. In P. G. Satinder (Ed.), *Cognition, Communication and Interaction: Transdisciplinary Perspectives on Interactive Technology* (pp. 114–129). London: Springer-Verlag.

Cohen, M. J. (2008). *Educating, counseling and healing with nature*. Project Nature Connect, Institute of Global Education.

Girardet, H. (2015). *Creating regenerative cities*. Routledge.

Henshilwood, C. S., d'Errico, F., van Niekerk, K. L., et al. (2018). An abstract drawing from the 73,000-year-old levels at Blombos Cave, South Africa. *Nature, 562*, 115–118. https://doi.org/10.1038/s41586-018-0514-3

Neis, H. J., Brown, G. A., Gurr, J. M., & Schmidt, J. A. (2012). Generative process, patterns, and the urban challenge. *Fall 2011 International PUARL Conference*. Portland, Oregon: PUARL Press.

PhysOrg. (2011). *Global resource consumption to triple by 2050: UN*. PhysOrg, 12 May 2011. Retrieved March 7, 2022 from https://m.phys.org/news/2011-05-global-resource-consumption-triple.html

Roös, P. B. (2021). *Regenerative-adaptive design for sustainable development—A pattern language approach* (Sustainable development goals series). Springer International. https://doi.org/10.1007/978-3-030-53234-5_1

Roser, M., Ritchie, H., & Ortiz-Ospina, E. (2019). World population growth. Published online at OurWorldInData.org. Retrieved September 3, 2022 from https://ourworldindata.org/world-population-growth. (Original work published 2013).

UN. (1992). *Agenda 21*. United Nations Conference on Environment and Development (UNCED), Rio de Janeiro, Brazil, June 3–14, 1992. Retrieved March 9, 2022 from https://sustainabledevelopment.un.org/outcomedocuments/agenda21

UN. (2015). *Transforming our world: The 2030 agenda for sustainable development*. A/RES/70/1 United Nations. Retrieved March 8, 2022 from https://sustainabledevelopment.un.org/content/documents/21252030%20Agenda%20for%20Sustainable%20Development%20web.pdf

UN. (2019). *Growing at a slower pace, world population is expected to reach 9.7 billion in 2050 and could peak at nearly 11 billion around 2100*, June 17, 2019. Department of Economic and Social Affairs, United Nations. Retrieved March 9, 2022 from https://www.un.org/development/desa/en/news/population/world-population-prospects-2019.html

# CHAPTER 2

# Beyond Sustainability: An Integral Framework

**Abstract** There is a desire worldwide to achieve sustainable outcomes and provide a better place to live for our global society. Many governments and organisations worldwide have set in place policies, value statements and visions, trying to instil a sustainable future for their citizens. However, there is a growing debate that sustainability is not good enough, and that as a global society we need to progress beyond sustainability—moving into the realms of regeneration and adaptation. There is a need for 'deep sustainability', and this chapter explores the consideration of an *Integral Biophilic Pattern Language Framework* to transition from a sustainable to a regenerative discourse, where the innate human-nature connection is positioned as critical in biophilic design and planning solutions.

**Keywords** Beyond sustainability • Regeneration • Regenerative-adaptive • Deep sustainability • Integral design • Biophilic patterns

© The Author(s), under exclusive license to Springer Nature
Switzerland AG 2022
P. B. Roös, *A Biophilic Pattern Language for Cities*, Sustainable
Urban Futures, https://doi.org/10.1007/978-3-031-19071-1_2

**Fig. 2.1**  Planetary wellbeing. (Image author: PB Roös, 2022)

## INTRODUCTION

The word 'sustainability' has become a word that is used on a daily basis by leading researchers, scientists, philosophers, environmentalists, socialists and government organisations, it can be said that it has become a 'household word' (Fig. 2.1). The aim of sustainability is to achieve a balance in a rapidly resource hungry world. The concept of sustainability, as defined by the Brundtland Report, has been described by Sim Van der Ryn and Cowan (1996) as bias towards anthropogenic and utilitarianism that is based on a technical approach for the better management of the non-human world for its continued existence (Van der Ryn & Cowan, 1996, p. 5). Under scrutiny of the definition of sustainability, it becomes clear that sustainable development seeks to minimise pollution and environmental impact rather than achieve regenerative results of natural ecosystems including clean air, water and soil (Roös, 2021).

There is a desire worldwide to achieve sustainable outcomes and provide a better place to live for our global society. Many governments and organisations worldwide have set in place policies, value statements and visions, trying to instil a sustainable future for their citizens. In 2015, the United Nations Member States adopted the *17 Sustainable Development Goals* (SDGs), claiming that these are the world's best plan to build a better world for people and the planet by 2030 (UN, 2019) (Fig. 2.2).

Van der Ryn and Cowan (1996) critique the common approach of sustainable development adopted on a global scale and argue that it is in essence a 'technological sustainability approach' with not enough

**Fig. 2.2** Sustainable Development Goals (by the United Nations, November 2019, public domain)

substance that can support both humans and other species, as well as to consider the values and cultures of a global society. It is interesting that in the early 90s various environmentalists argued that ecological sustainability is paramount and needs to include cultural values. A key proponent of the 'ecological sustainability' approach was David Orr (1992). He argued that if ecological sustainability is applied to solutions for a resilient and ecological future, it has a very different vision of outcomes for society than the standard 'sustainability' approach:

> Ecological sustainability is the task of finding alternatives to the practices that got us into trouble in the first place; it is necessary to rethink agriculture, shelter, energy use, urban design, transportation, economics, community patterns, resource use, forestry, transportation, the importance of wilderness and our cultural values (Orr, 1992, p. 24).

This approach by Orr provides the fundamentals for the considerations of regeneration, and as argued by Christopher Alexander (2004), the considerations need to be embedded within a 'deep sustainability' that addresses the core of humanity's global culture, the planet's global environment and human's deep connection to nature (Roös, 2021, p. 8).

*Deep sustainability*[1] as defined by Alexander can be considered to be in contrast with the general understanding and current practice of sustainability (Alexander, 2004). In the overall perspective of sustainability, along with the application of the triple bottom line to implement sustainability, the natural environment is seen as a resource provider—in essence, in the decision-making process of a global economic system it is how to maintain a constant pool of resources into the future for ongoing human use and benefits (O'Riordan, 2009). In contrast, 'deep sustainability' is the consideration of the whole ecological system together, where humans are part of nature, and nature is collectively one system. Deep sustainability considers the benefits to the human and the more-than-human collectively. To achieve deep sustainability, an 'integral' approach is necessary to deal with the complexities of this concept.

---

[1] *Deep sustainability* goes beyond the mere aspects of resource efficiency, energy reduction or sustainable growth; it requires the emotional, spiritual and cultural connections of people with their built and natural environment. Deep sustainability addresses the 'ends', rather than the means to an end (Alexander, 2004, p. 6).

## INTEGRAL SUSTAINABLE DESIGN FRAMEWORK

When considering progressing beyond sustainability and transition into a regenerative state of affairs, we will need a new 'integral' approach to the way we design and build our cities. The integral approach to sustainability is found in the consideration of multilevel complexity, the intersecting of the domains of self, culture and nature. If global society wishes to put nature first and wants to achieve a resilient and sustainable future, it is time to rethink the fundamentals of sustainable development and design. In *Integral Sustainable Design: Transformative Perspectives* DeKay (2011), Mark DeKay proposes an integral theory of sustainable design that moves from a standard approach to an integral model. The *Integral Theory* as a meta-theory organises its fundamentals in two contexts (2011, p. xxiii):

> *The four perspectives*, which arise from fundamental distinctions of value found in language (I, We, It/Its) and represent the methods of arts, humanities, basic sciences and the complex sciences… and
> *Levels of complexity*, which arise from the unfolding sequence of development in human individuals, cultures and physical systems, which manifest as developmental sequences such as those values, cognition, biological evolution, economic systems and worldviews (DeKay, 2011, p. xxiii).

*Integral Sustainable Design* goes beyond the basic requirements for sustainability, and investigates the achievement of sustainability from a technological, ecological, experiential and cultural context. Dekay's integral sustainable design theory views sustainability from four perspectives (DeKay, 2011, p. xxiv):

- Perspective of *behaviours*—the *what* of individual parts.
- Perspective of *systems*—the *how* of complex wholes.
- Perspective of *experiences*—the *who* that intends, thinks and feels.
- Perspective of *cultures*—the *why* of the collective we.

The *Integral Theory* by Wilber (2001) combined with the principles of sustainable design unites the philosophy, arts and ethics of design with science. From the four-quadrants perspective, *Integral Sustainable Design* consists of:

- *Patterns of form* that order social and ecological relationships.
- With individual members and parts including their activities, functions and *performance*.
- *Experiences* of systemic members including perception, sentience and awareness.
- Various *levels of complexity* from individual members acting with each other sharing meaning and understanding.

When applying this four-quadrant perspective to sustainable development, a deeper sustainable outcome emerges. As indicated in Fig. 2.3, sustainability (achieved in sustainable development)—as part of our natural environment is viewed as *experiences*—self and conscious; *behaviours*—science and performance; *cultures*—meaning, worldviews and symbolism; and *systems*—social and natural ecologies (DeKay, 2011, p. 17).

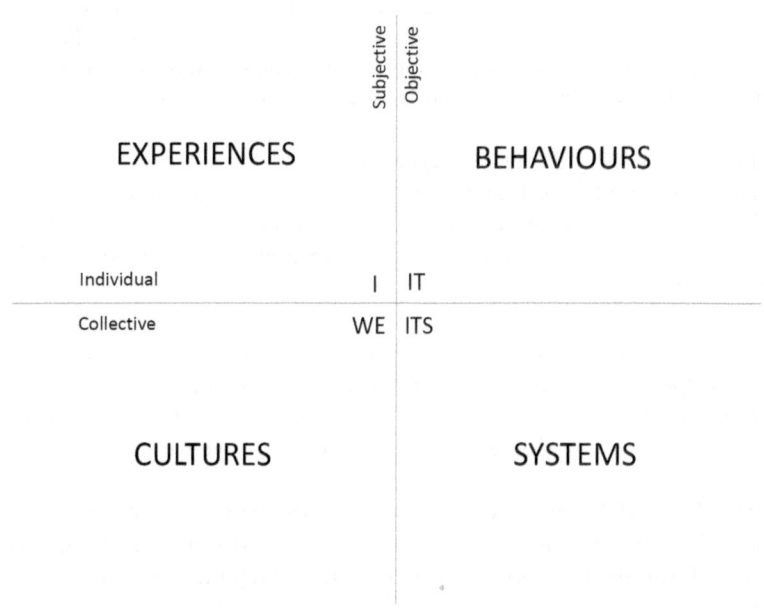

**Fig. 2.3**  Sustainable design perspectives embedded in the integral theory. (Image author: PB Roös, 2022, adapted from DeKay, 2011)

What is important to note—is that any decision or framework considering a sustainable future—needs to accept the interconnectedness of humans and nature. The planning and design of sustainable cities should go beyond structures, infrastructure and buildings; it must go to the heart of the life-giving phenomena of this planet—natural systems. Using the systems of nature and the knowledge that nature provides, we can plan and design our built environments to mimic nature. This process starts with planning at ecological levels of scale to *design with nature* (McHarg, 1992/1969), and progressing to the deeper understanding of, and design with *living structures* through the nature of order—*a pattern language* (Alexander et al., 1977; Alexander, 2001–2005). It is these *living structures* in nature that trigger a neurological response in us as humans, a response that has a positive psychological effect. This exposure to a biophilic environment has been documented in classic experiments by Roger Ulrich (1984), and subsequently 'healing environments' and various other beneficial effects have been studied and recorded, similar to the biophilic effect (Salingaros, 2015; Rakel et al., 2018; von Lindern et al., 2016; Iyendo & Alibaba, 2014; Huisman et al., 2012).

It is my argument that if we want a true sustainable future, let me reframe—a regenerative and resilient future—we need to return to first principles, that is our innate human affiliation with the natural world—*biophilia*. Biophilia positioned at the four integral perspectives requires a framework that is an *Integral Biophilic Pattern Language Framework*, effectively a framework that can support creating liveable and healthy cities, a framework that is the foundation for going beyond sustainability, resulting in regeneration and resilience—both for humans and the more-than-human.

## AN INTEGRAL BIOPHILIC PATTERN LANGUAGE FRAMEWORK

Consider the four perspectives of the integral sustainable design theory as noted earlier, and to frame the application of these perspectives in relation to nature and the human-nature connection in the design of our built environments, a framework that puts nature (bio) first, results in a framework that considers nature experiences, nature elements, nature contexts and nature meanings. This supports the all-encompassing principle of 'integral sustainability' from the technological, ecological, experiential and cultural context. This integral thinking embedded in ecological consciousness, requires us to transcend to a worldview that embraces the four integral perspectives of (1) the interior individual, the self, the 'I'; (2) the exterior individual, the physical, the behaviour, the 'It'; (3) the exterior collective, systems, the 'Its'; and (4) the interior collective, the values, the cultures, the meanings, the 'We' (Roös, 2022, p. 45). This all-inclusive awareness of the perspectives is then categorised in the '*Integral Biophilic Pattern Language Framework*' (Fig. 2.4), and consists of the following:

- *Biophilia*—Love of nature, love of all living beings, conscious and unconscious, deep connections to nature, subjective nature, self, self-awareness, truthfulness, sincerity.
- *Biologic*—Empirical forms, objective nature, environmental performance, living objects.
- *Biosystem*—Functional fit, evolutionary adaptation, natural systems, social systems, living interactions, inter-objective nature.
- *Bioculture*—Nature understanding, nature meanings, ethics and morals, rightness, justness, cultural fit, inter-subjective.

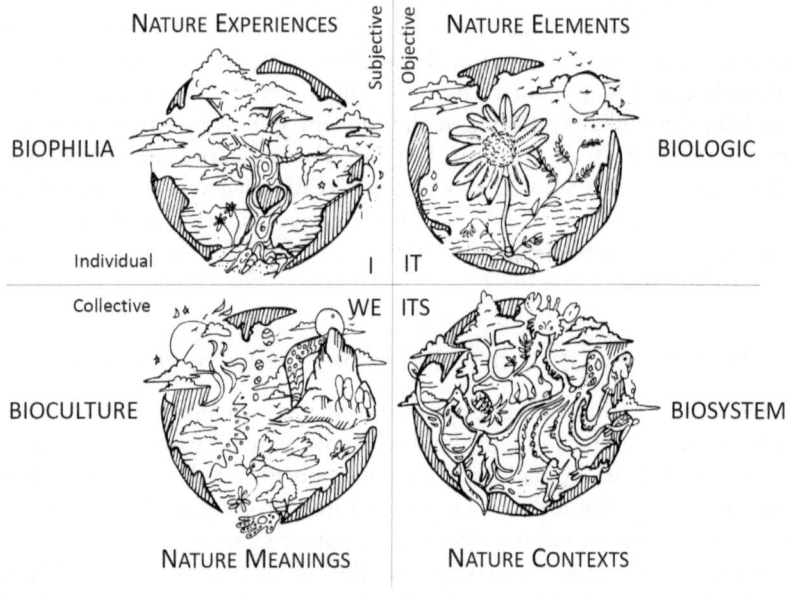

**Fig. 2.4** Integral biophilic pattern language framework. (Image author: PB Roös, 2022)

In developing a *Biophilic Pattern Language for Cities*, each perspective and relevant attribute in the four integral quadrants are aligned with four meta biophilic patterns namely:

DIRECT EXPERIENCE OF NATURE [1][2]
INDIRECT EXPERIENCE OF NATURE [2]
EXPERIENCE OF PLACE, SPACE AND ATTACHMENT [3]
NATURE PATTERNS, PROCESSES AND SYSTEMS [4]

Each biophilic pattern is used to provide guidance for design decisions and options in the planning of cities, considering the inclusion of nature and the innate human-nature connection as a critical biophilic design criteria, further described in detail in the following chapters of this book.

## REFERENCES

Alexander, C. (2001–2005). *The nature of order—An essay on the art of building and the nature of the universe, Book One: The phenomenon of life*. The Center for Environmental Structure.

Alexander, C. (2004). *Sustainability and morphogenesis: The birth of a living world*. Centre for Environmental Structure.

Alexander, C., Ishikawa, S., Silverstein, M., Jacobson, M., Fiksdahl-King, I., & Angel, S. (1977). *A pattern language: Towns, buildings, construction*. Oxford University Press.

DeKay, M. (2011). *Integral sustainable design: Transformative perspectives*. Earthscan.

Huisman, E. C., Morales, E., van Hoof, J., & Kort, H. S. (2012). Healing environment: A review of the impact of physical environmental factors on users. *Building and Environment, 58*, 70–80. Retrieved from https://core.ac.uk/download/pdf/82518574.pdf

Iyendo, T. O., & Alibaba, H. Z. (2014). Enhancing the hospital healing environment through art and day-lighting for user's therapeutic process. *International Journal of Arts and Commerce, 3*(9), 101–119. Retrieved from https://www.researchgate.net/publication/306394295

McHarg, I. (1992/1969). *Design with nature*. John Wiley & Sons.

O'Riordan, T. (2009). Reflection on the pathways of sustainability. In W. N. Adger & A. Jordan (Eds.), *Governing sustainability* (pp. 307–328). Cambridge University Press.

---

[2] Note the patterns are written in 'small caps', as to follow the standard writing style of pattern names as per Christopher Alexander's A Pattern Language (1977).

Orr, D. W. (1992). *Ecological literacy: Education and the transition to a postmodern world.* Albany State University of New York.

Rakel, D., Sakallaris, B. R., & Jonas, W. (2018). Chapter 2: Creating optimal healing environments. In D. Rakel (Ed.), *Integrative medicine* (4th ed., pp. 12–19). Elsevier. https://doi.org/10.1016/B978-0-323-35868-2.00002-5

Roös, P. B. (2021). *Regenerative-adaptive design for sustainable development—A pattern language approach* (Sustainable development goals series). Springer International. https://doi.org/10.1007/978-3-030-53234-5_1

Salingaros, N. A. (2015). *Biophilia and healing environments.* Terrapin Bright Green LLC and Levellers Press. Retrieved from https://www.terrapinbrightgreen.com/wp-content/uploads/2015/10/Biophilia-Healing-Environments-Salingaros-p.pdf

UN. (2019). *Helping governments and stakeholders make the SDGs reality.* United Nations Sustainable Development Goals Knowledge Platform. Retrieved September 16, 2019 from UN—SDGs Platform: https://sustainabledevelopment.un.org

Van der Ryn, S., & Cowan, S. (1996). *Ecological design.* Island Press.

von Lindern, E., Lymeus, F., & Hartig, T. (2016). The restorative environment: A complementary concept for salutogenesis studies. In M. Mittelmark et al. (Eds.), *The handbook of salutogenesis* (pp. 181–195). Springer. https://doi.org/10.1007/978-3-319-04600-6_19

Wilber, K. (2001). *A theory of everything: An integral vision for business, politics, science, and spirituality.* 1st Ed. Massachusetts: Shambala Publications, Inc.

# CHAPTER 3

# Direct Experience of Nature [Pattern 1]

**Abstract** Edward O. Wilson noted in his Biophilia Hypothesis (1986) that humans need daily contact with the natural world to be healthy and experience wellbeing. The biophilia theory supports the argument that human habitated environments need to include elements of nature to provide us with psychological and physiological health. In highly densified cities made of tarmac, concrete, steel and glass with increasingly high-rise buildings and skyscrapers, it is almost impossible to have daily contact with nature. It is modern architecture and these buildings that are at the core of unhealthy and non-sustainable built environments. This chapter designates the first meta biophilic pattern—DIRECT EXPERIENCE OF NATURE [1]—to inject nature back into cities.

**Keywords** Biophilia • Human-nature connection • Nature experience • Nature in cities • Sustainable habitats • Biophilic urbanism

© The Author(s), under exclusive license to Springer Nature Switzerland AG 2022
P. B. Roös, *A Biophilic Pattern Language for Cities*, Sustainable Urban Futures, https://doi.org/10.1007/978-3-031-19071-1_3

**Fig. 3.1** Direct experience of nature in the Otway forest. (Image author: PB Roös, 2022)

## Introduction

The human neurological reactions to biophilic environments have a positive psychological and physiological effect (Fig. 3.1), evidenced in a raft of literature presenting claims that biophilia results in health and wellbeing advantages, as a result of experimental outcomes as cited by Salingaros

(2019) (including Kellert, 2018; Ryan & Browning, 2018; Salingaros, 2015; Kellert et al., 2008; Joye, 2007).

Edward O. Wilson noted in his Biophilia Hypothesis (1986) that humans need daily contact with the natural world to be healthy and experience wellbeing. His proposition was that humans have co-evolved with nature and that biophilia is "the innately emotional affiliation of human beings to other living organisms. Innate means hereditary, and hence biophilia is part of ultimate human nature" (Kellert & Wilson, 1993, p. 31). The biophilia theory supports the argument that our environments of the habitats of humans need to include the elements of nature to provide us with psychological and physiological health:

> Over thousands of generations the mind evolved within a ripening culture, creating itself out of symbols and tools, and genetic advantage accrued from planned modifications of the environment. The unique operations of the brain are the result of natural selection operating through the filter of culture. They have suspended us between the two antipodal ideas of nature and machine, forest and city, the natural and artificial, relentlessly seeking, in the words of the geographer Yi-Fu Tuan, an equilibrium not of this world (Wilson, 1984, p. 12).

This deep affiliation with the natural environment has been recognised by Alexander as noted in *A Pattern Language* (1977), described in Pattern 173—Garden Wall:

> People need contact with trees and plants and water. In some way, which is hard to express, people are able to be more whole in the presence of nature, are able to go deeper into themselves, and are somehow able to draw sustaining energy from the life of plants and trees and water (Alexander et al., 1977, p. 806).

Several propositions by various authors identify the importance of biophilia and the application of the biophilia hypothesis to design practice. *Nature informed design* as an architectural practice was proposed by Kellert et al. (2008) in *Biophilic Design*, and city planning by Beatley (2011) in *Biophilic Cities*, involved frameworks that offer an approach that reconnects people with nature. What is noticeable in the literature, is that the 'direct connection to nature', or 'direct experience of nature', is fundamental in human physiological and psychological wellbeing (Roös, 2021; Salingaros, 2019; Kellert, 2018; Kellert et al., 2008). A balanced nature and human ecosystem supported by biophilic design is the foundation for liveable cities. Liveable and healthy cities result in sustainable

cities, which in turn supports healthy ecosystems and natural environments. It is thus imperative to create built environments in cities that
includes spaces that enhances the DIRECT EXPERIENCE OF NATURE [1].

## PATTERN STATEMENT

*People need direct contact with nature, experiencing the basic qualities and
characteristics of the natural environment. Healthy habitats for humans in
cities need to include naturalistic features such as plants and trees, animals,
landscapes, sunlight, fresh air, water, weather, and views of nature.*

## DISCUSSION

Research has shown that the affiliation with nature continues to be critical
in modern-day human health and wellbeing (Kellert, 1997, 2012), and
has been strongly identified as a crucial issue in the health sciences. The
exposure to nature continues to result in wellbeing in a wide range of settings; at work, home, recreation, community areas and within the city
environments (Roös & Jones, 2017; Kellert, 2012, Browning et al.,
2014). The benefits from contact with nature depend on repeated experience of the biophilic effect (Roös et al., 2018), yet the majority of spaces
in cities are harsh, unnatural, monotonous and without character, lifeless.

Many people recognise the benefits of vegetation and gardens, and
experience a sense of wellbeing when walking or spending time in these
natural environments. The result of biophilia—or the biophilic effect—
happens when two distinct mechanisms are present (Mehaffy et al., 2020;
Salingaros, 2019):

1. An intimate contact with other living beings, such as plants, animals,
   and people.
2. Response to geometries (and elements) that are created by biological rules (and the environment).

It is worth noting that I have included 'elements' and 'environment' as
part of the distinct mechanisms to achieve a biophilic effect, for the simple
reason that as humans our biological responses to our surroundings are
grounded in the fact that all of our senses work together at any given time.
This is particularly evident when we walk in a forest, the sensory phenomena engaging us is heightened with what we see, feel, touch, smell, hear
and even the inclusion of unconscious alertness.

Biophilic design can be integrated into the built environment, using these rules of creating a 'biophilic effect' as a DIRECT EXPERIENCE OF NATURE [1], stimulating the human senses. The senses triggered are a basic characteristic of how the biophilic attributes are experienced. The biophilic attributes include plants and trees, animals, landscapes, sunlight, air, water, weather and views of nature, further described in detail below (Fig. 3.2).

**Fig. 3.2** Biophilic attributes evident in an urban setting. (Image author: PB Roös, 2022)

## BIOPHILIC ATTRIBUTES

The biophilic attributes resulting in a DIRECT EXPERIENCE OF NATURE [1] can broadly be described as environmental features. The biophilic effect is a sensory response to direct environmental conditions, that includes the following attributes, adapted from Kellert (Kellert, 2018; Kellert et al., 2008):

1. *Plants and trees*—are probably the most important attribute to integrate into the built environment to create direct contact between city dwellers and nature. Deliberate vegetation design must identify native species to support biodiversity and ecologically intact ecosystems, climate resilience and naturalistic elements.
2. *Animals*—are deeply embedded in human biology and support the biophilia hypothesis of innate attraction to living beings. Identify the local animals and create habitats such as gardens, nature strips, roof gardens and parklands that will attract local birds and animals to support biodiversity corridors of the area.
3. *Landscapes*—that possess a close relationship to local habitats and ecosystems support a sustainable city. Typical landscapes that enhance human wellbeing, due to their importance during human evolution, include distributed shrubs and trees, grasslands, colourful foliage, different shades of green, flowers, prospect views, water and savannah-like settings.
4. *Sunlight*—fundamental ingredient for life on this planet, natural light is one of the most basic aspects of human and more-than-human existence. The experience of sunlight effects how humans respond to their environment, orient themselves, and relate to diurnal changes, daylight shifts and seasonal patterns. Sunlight also has therapeutic qualities. Deliberate design that considers sunlight penetration into streets, alleys and buildings can provide ambient qualities that stimulate interest, knowledge of space, awareness, circadian rhythms, and increase health.
5. *Fresh air*—another fundamental quality of the environment that is necessary for life. Although air is invisible, it is one of our human basic needs, or as per Cohen (2008), one of our senses that is an unconscious desire to breathe. Without oxygen—fresh air—no life can exist. The qualities of air around us are of utmost importance, and considering the high pollution levels in cities, more so fresh air.

Increasing natural ventilation, access to outdoor environments, and manipulating basic atmospheric conditions such as airflow, barometric pressure, temperature, and humidity through natural processes can enhance wellbeing.

6. *Water*—the exposure to water can result in significant physical and psychological benefits, including stress relief, enhanced performance, enhanced creativity, and increased feelings of tranquillity. Presence of water can transfer dull environments to stimulating spaces inclusive of constructed waterfalls, fountains, rainwater spouts, aquaria, and ponds. Further, the presence of water attracts animals and increases biodiversity within cities.

7. *Weather*—provides a connection to a specific place by the experience of sunshine, wind, rain, temperature and other climatic conditions. Weather informs us of daily changes in temperature, as well as the changes of seasons. Knowing the weather and seasonal changes allows us to grow produce and be engaged in gardening both at home and in community settings. Purposefully designed places that allow engagement with the weather, including porches, terraces, courtyards and roof gardens will result in both a closer contact with the outdoors, as well as an increased awareness of our independence of the natural world.

8. *Views of nature*—enrich a sense of direct contact between humans and nature. A view to an element of nature, living systems, natural processes and landscapes generally exert the greatest impact when they are at a relatively moderate to short distance, and from a sheltered setting. Prospect—an unimpeded view over an area or distance allows for surveillance, and results in a sense of security and belonging (Fig. 3.3).

## Pattern Applications

There are various possibilities to integrate the biophilic attributes of the pattern Direct Experience of Nature [1] into the urban environment (Fig. 3.4). Provide setbacks in each parcel and city block to allow for front gardens and trees, include trees in the road reserve, include a minimum of 40% of surface coverage with *Plants and Trees* to assist in city greenery,

**Fig. 3.3** Direct view to nature with prospect. (Image author: PB Roös, 2022)

biodiversity and the mitigation of urban heat and climate impacts. Include roof gardens, courtyard gardens, vertical green living walls and façades, public green spaces, community gardens and urban vegetable gardens. Use native plants and trees to attract local *Animals*, supporting a

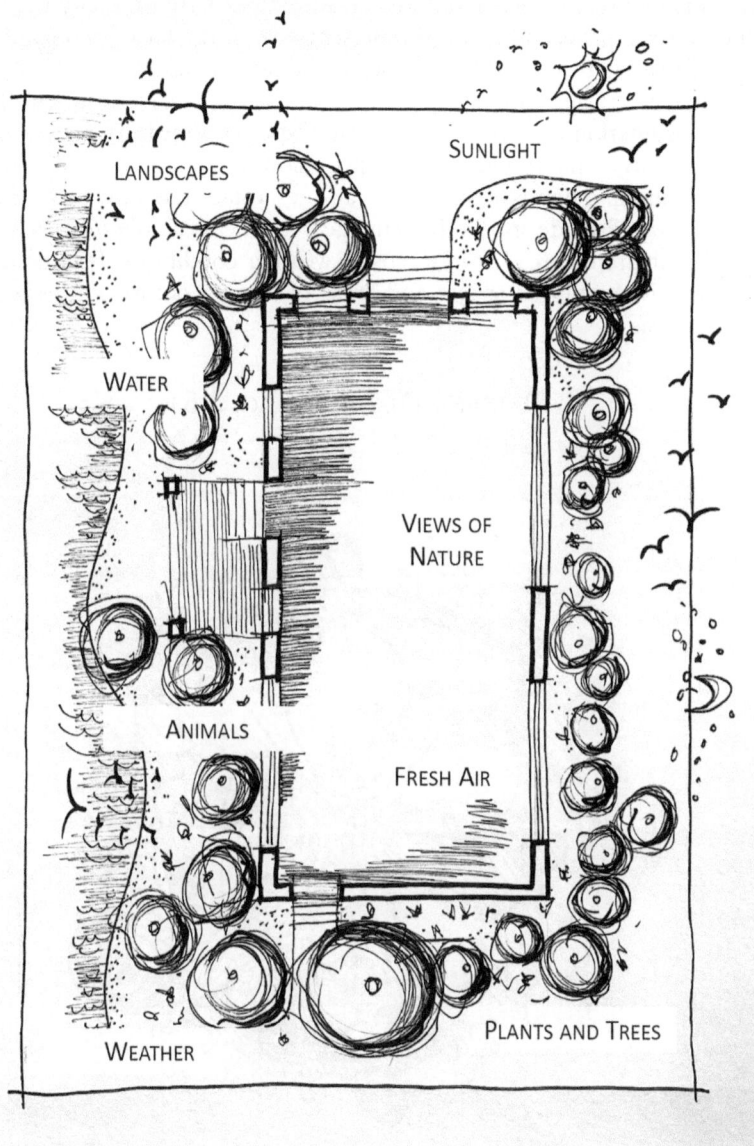

**Fig. 3.4** Application diagram of biophilic attributes for DIRECT EXPERIENCE OF NATURE [1]. (Image author: PB Roös, 2022)

sustainable urban ecosystem. Create *Landscapes* within the city by intro-
ducing large parks and gardens representative of the local ecology. Avoid
out-of-scale high-rise buildings and introduce setbacks to allow *Sunlight*
into the streetscapes and occupied building complexes. Introduce in abundance vegetation and trees into the cityscapes as a measure for pollution
control, natural filters and *Fresh Air* provision. Provide water features such
as fountains and ponds accessible for all ages to facilitate direct connection
with *Water*. Provide spaces that engage with the local *Weather*, as well as
provide sheltered areas from intensive weather conditions. Include direct
*Views of Nature* in all spaces and areas of human occupation as well as
public areas.

PATTERN DIAGRAM (FIG. 3.5)

**Fig. 3.5** Sketch of pattern—DIRECT EXPERIENCE OF NATURE [1]. (Image author:
PB Roös, 2022)

## PATTERN LINKS

### *Upward Links*

Embedded in a hierarchal structure within a larger network of patterns, the biophilic pattern DIRECT EXPERIENCE OF NATURE [1] links upwards to patterns in a higher level of the hierarchical scale—the fundamental patterns—of the *Regenerative-Adaptive Pattern Language* (Roös, 2021), to support an overall healthy planet as indicated in the pattern—THE WHOLE [P1], human connections to nature—LOVE OF NATURE [P6], and actions that result in positive outcomes—NATURE'S DESIGN [P7].[1]

### *Downward Links*

Although the pattern DIRECT EXPERIENCE OF NATURE [1] involves actual contact with real attributes of nature, this experience only represents a starting point in engaging with nature in the built environment (Kellert, 2018, p. 24). In the built environment biophilic attributes as images or representations of nature result in an INDIRECT EXPERIENCE OF NATURE [2]. This pattern draws on the unique human capacity to convert objective and empirical reality into metaphorical forms and symbolic representations of nature.

### *Integral Framework Alignment*

The attributes of the pattern DIRECT EXPERIENCE OF NATURE [1] aligns mostly with the Integral Framework's upper left quadrant: *Nature Experiences—Biophilia*, the individual person's direct experience referred to the 'I'. This perspective includes the actual experiences from the connections to these elements in direct connection to nature.

However, due to shapes and forms and complexity in these natural elements, as well as actual physical elements of direct connection, to a degree the attributes also align with the upper right quadrant: *Nature Elements—Biologic*.

---

[1] The relationship of the '*biophilic patterns*' with the higher level '*fundamental patterns*' of the *Regenerative-Adaptive Pattern Language* are further described in Chap. 7.

## Comparison/Relationship with Patterns and Metrics by Other Authors

Various literature by other authors attempts to describe patterns, attributes and different metrics to include biophilic design elements in the built environment (Mehaffy et al., 2020; Salingaros, 2019; Kellert, 2018; Kellert et al., 2008; Browning et al., 2014; Alexander et al., 1977). This can be overwhelming, and in some instances confusing. In this book I endeavour to align different patterns and attributes with the four meta biophilic patterns described, as well as provide some examples of similar patterns.

For example, the pattern *Biophilic Urbanism* [2.4] by Mehaffy et al. (2020) summarises various biophilic patterns and attributes relating to the statement that people have an instinctive need to be surrounded by the forms of nature, including biologic nature (p. 57). The pattern notes the ten factors of biophilic design as per Salingaros (2019) and reference the 'biophilic index'. However, the chapter does not go into the detailed descriptions of each factor to allow understandings on how to apply these ten factors. In contrary, the patterns by Browning et al. (2014) describe in more detail the context of each pattern, its application and the possible health and wellbeing, or biological responses to these patterns (Browning et al., 2014, pp. 23–51). The patterns Visual Connection to Nature [1], Non-visual Connection to Nature [2], Thermal Airflow, Variability [4], Presence of Water [5], and Dynamic and Diffuse Light [6] all include attributes that provide a direct contact with natural elements, and according to Browning et al., these are categorised as the elements of the "Nature in the Space" (Browning et al., 2014, p. 9). Nature in the space addresses the direct, physical and ephemeral presence of natural elements in the space. An example of this condition is pattern 1—Visual Connection with Nature—that represents a direct view to elements of nature, living systems and natural processes. This is an example of *one element* that contributes to the biophilic pattern DIRECT EXPERIENCE OF NATURE [1]. As such, Table 3.1 attempts to align the pattern DIRECT EXPERIENCE OF NATURE [1] with other similar patterns and metrics, to assist the designer in the application of these.

**Table 3.1**  Pattern comparison/relationship with others: DIRECT EXPERIENCE OF NATURE [1]

| | |
|---|---|
| Roös (2022) | Direct Experience of Nature [1][a] |
| Mehaffy et al. (2020) | Biophilic Urbanism [2.4] |
| | Urban Greenway [3.1] |
| | Street Trees [8.3] |
| Salingaros (2019) | Sunlight |
| | Water |
| | Life |
| Kellert et al. (2008), Kellert (2018) | Direct Experience of Nature |
| | Environmental Features, Light and Space |
| Terrapin (Browning et al., 2014) | Visual Connection to Nature [1] |
| | Non-visual Connection to Nature [2] |
| | Thermal Airflow Variability [4] |
| | Presence of Water [5] |
| | Dynamic and Diffuse Light [6] |
| Alexander et al. (1977) | Green Streets [51] |
| | Access to Water [25] |
| | Pools and Streams [64] |
| | Still Water [71] |
| | Wings of Light [107] |
| | Indoor Sunlight [128] |
| | Light on Two Sides of a Room [159] |
| | Sunny Place [161] |
| | Roof Garden [118] |
| | Climbing Plants [246] |

[a] [1] indicates a *pattern*. When not indicated with [ ] the description in the table refers to an *attribute* or a *metric*

## CONCLUSION STATEMENT

We know that the future habitation of humans is urban. Urbanisation is one of the biggest mega trends of our time, and this trend shows that human settlements of villages become towns, towns turn into cities and cities become megacities (Roös, 2021, p. 19). Further, this trend shows that sprawling cities are made of tarmac, concrete, steel and glass (Girardet, 2015, p. 3). It is this modern architecture that is at the core of unhealthy and non-sustainable built environments. A *Biophilic Pattern Language for Cities* aims to address the issue of unhealthy compact city spaces, and the first pattern to inject nature back into cities is the DIRECT EXPERIENCE OF NATURE [1].

*Therefore:*

*If we want to move beyond damaging, destructive, unhealthy and unsustainable habitats for humans, and the more-than-human in cities, transitioning to liveable and healthy cities, we need to design and plan spaces that include the biophilic attributes of plants and trees, animals, landscapes, sunlight, air, water, weather, and views of nature.*

## REFERENCES

Alexander, C., Ishikawa, S., Silverstein, M., Jacobson, M., Fiksdahl-King, I., & Angel, S. (1977). *A pattern language: Towns, buildings, construction*. Oxford University Press.

Beatley, T. (2011). *Biophilic cities*. Washington: Island Press.

Browning, W. D., Ryan, C. O., & Clancy, J. O. (2014). *14 Patterns of biophilic design*. Terrapin Bright Green, LLC.

Cohen, M. J. (2008). *Educating, counseling and healing with nature*. Project Nature Connect, Institute of Global Education.

Girardet, H. (2015). *Creating regenerative cities*. Routledge.

Joye, Y. (2007). Architectural lessons from environmental psychology: The case of biophilic architecture. *Review of General Psychology, 11*(4), 305–328. https://doi.org/10.1037/1089-2680.11.4.305

Kellert, S. (1997). *Kinship to mastery: Biophilia in human evolution and development*. Island Press.

Kellert, S. (2012). *Birthright: People and nature in the modern world*. Yale University Press.

Kellert, S., & Wilson, E. O. (1993). *The biophilia hypothesis*. Island Press.

Kellert, S. R. (2018). *Nature by design*. Yale University Press.

Kellert, S. R., Heerwagen, J. H., & Mador, M. L. (Eds.). (2008). *Biophilic design. The theory, science and practice of bringing buildings to life*. Wiley.

Mehaffy, M. W., Salingaros, N. A., Kryazheva, Y., & Rudd, A. (2020). *A new pattern language for growing regions*. Sustasis Press.

Roös, P. B. (2021). *Regenerative-adaptive design for sustainable development—A pattern language approach* (Sustainable development goals series). Springer International. https://doi.org/10.1007/978-3-030-53234-5_1

Roös, P. B., & Jones, D. (2017). Knowledge of making life: Design patterns for regenerative-adaptive design. In P. K. Collins (Ed.), *DesTech 2016: Proceedings of the International Conference on Design and Technology* (pp. 203–210). Knowledge E. https://doi.org/10.18502/keg.v2i2.616

Roös, P. B., Jones, D. S., Downton, P. B., & Zeunert, J. (2018). Biophilic railway stations: Re-imagine the nature of transit design. In IFLA 2018: Biophilic City, Smart Nation, and Future Resilience: Proceedings of the 55th International Federation of Landscape Architects World Congress 2018 (pp. 800–813). International Federation of Landscape Architects.

Ryan, C. O., & Browning, W. D. (2018). Biophilic design. In R. A. Meyers (Ed.), *Encyclopedia of sustainability science and technology* (pp. 1–44). Springer. https://doi.org/10.1007/978-1-4939-2493-6_1034-1

Salingaros, N. A. (2015). *Biophilia and healing environments.* Terrapin Bright Green LLC, Levellers Press. Retrieved from https://www.terrapinbrightgreen. com/wp-content/uploads/2015/10/Biophilia-Healing-Environments-Salingaros-p.pdf

Salingaros, N. A. (2019). The biophilic index predicts healing effects of the built enviroment. *Journal of Biourbanism, 8*(1), 13–34.

Wilson, E. O. (1984). *Biophilia.* Harvard University Press.

# CHAPTER 4

# Indirect Experience of Nature [Pattern 2]

**Abstract** Considering the impact of nature, human beings are sensory beings constantly engaged with the patterns of nature, and other environmental phenomena that exist in the environments they inhabit. Many patterns of nature are available to trigger and work with our senses, both directly and indirectly—consciously and unconsciously. This 'biophilic effect' is crucial for our physiological and psychological health and wellbeing. Humans are inherently part of the natural world, even in a modern industrial society ignorant of nature with a focus on technology and manufacturing, this innate human-nature connection continues to shape our evolution as a species, and undeniably exists. This chapter is designated to the second meta biophilic pattern—INDIRECT EXPERIENCE OF NATURE [2]—to guide the sustainable and healthy cities agenda to include the characteristics of a complex order derived from nature, the living structures.

**Keywords** Living structures • Biophilic effect • Human-nature affiliation • Healthy cities • Sustainable cities • Health and wellbeing

P. B. Roös, *A Biophilic Pattern Language for Cities*, Sustainable Urban Futures, https://doi.org/10.1007/978-3-031-19071-1_4

**Fig. 4.1**    Fractal geometry and complexity. (Image author: PB Roös, 2022)

## Introduction

All animals, including humans, need to know and process the information from their surrounding environment for basic survival (Fig. 4.1). This information acquired from observing the environment through sensory inputs, is first rapidly encoded as patterns inherent in the functioning of the brain (Sweeny et al., 2011; Wang et al., 2009). The large numbers of encoded images and sound patterns from the surrounding environment are then recalled and mentally manipulated in ways that enable comparisons of different patterns, helping to assist in making basic decisions. Only experienced in the human brain, the generation of new patterns can convey objects and processes that could possibly exist, even though they are not directly visible at the time of observation (Mattson, 2014).

By nature, human beings are sensory beings, constantly engaging with patterns of nature and other environmental phenomena that exist in the environments that they inhabit. Many patterns of nature are available to trigger and work with our senses, both directly and indirectly—consciously and unconsciously. This is evident when we engage with nature, and distinct natural attraction senses and sensitivities are triggered. Nature is multi-sensory, informed by our conscious and unconscious innate connection to it (Roös, 2021, p. 82). As noted in the previous chapter, two distinct mechanisms trigger the 'biophilic effect', one—the direct contact with nature and living beings, and two—the response to geometries or representations of structure and elements in nature. Salingaros (2019) argues that "the complex geometry of the environment is responsible for the biophilic effect, but it has to be a special type of complexity" (Salingaros, 2019, p. 3). This geometry is fundamental for biophilic design, and Salingaros further identified that "representations-of-nature" and "organised-complexity" need to be part of a design practice to be able to create healing built environments (Salingaros, 2019, p. 4).

However, the biophilic effect is a consequence from our biological responses to the natural environment (Murchie, 1978), and this effect of *biophilic patterns* is a result of many *multi-sensory phenomena*, when experienced first-hand, is difficult to describe in words (Roös, 2021; Cohen, 2008).

The pattern INDIRECT EXPERIENCE OF NATURE [2] relies on these phenomena of geometries, forms, shapes, complexities and sensory contexts directly representative of the natural world. This indirect experience is embedded in the unique human capacity to "convert empirical and objective reality into symbolic and metaphorical forms through projecting thoughts, images and feelings" (Kellert, 2018, p. 24).

## PATTERN STATEMENT

*People have an instinctive need to be connected consciously and unconsciously to the living structures of the natural world, inclusive of natural shapes, forms, patterns and processes.*

## Discussion

An extensive amount of research has shown that the inclusion of natural characteristics in the built environment supports the enhancement of human health and wellbeing (Mehaffy et al., 2020). This increased wellbeing is due to our innate affiliation with the natural world, through the phenomena of *biophilia*. In the built environment it is possible to extend this innate connection with the natural world, to the structures, buildings and public spaces in cities. Examples of traditional buildings give us inspiration to include in the design of current cityscapes the geometries, fractal and ornamented elements representative of nature. The example of historical architecture shows us that organised complexity integrated into the building structure results in old-fashioned ornamental façades, roof structures, entrances and internal environments (Salingaros, 2019).

These buildings provide to a certain level the elements needed for the biophilic effect, however without plants and gardens, the desired result is incomplete. Good building design and cityscape design that achieve intimate and direct contact with real nature trigger enhanced positive emotions from the close interaction with living plants (Salingaros, 2019; Dravigne et al., 2008; Brethour et al., 2007). It is thus important to always combine the criteria of both patterns the DIRECT EXPERIENCE OF NATURE [1], and INDIRECT EXPERIENCE OF NATURE [2], in design and planning solutions (Fig. 4.2).

## Biophilic Attributes

The biophilic attributes resulting in an INDIRECT EXPERIENCE OF NATURE [2] can broadly be described as those elements that unconsciously trigger an emotional affiliation with the natural world. The biophilic effect in this instance is as a result of natural shapes, forms, patterns and processes, included in the following attributes:

1. *Botanical motifs*—represent shapes and forms directly derived from plants such as flowers, leaves, stems, vines and trees.
2. *Biomorphic ornament*—shapes that are reminiscent of living organisms in rounded and irregular abstract organic forms.

**Fig. 4.2** Biophilic elements evident in ornament. (Image author: PB Roös, 2022)

3. *Natural geometry and patterns*—representation of elements in nature that have geometric repetitive qualities and patterns.
4. *Material and texture*—finishes or materials used that directly represent the textures of nature including stone, soils, clay and wood.
5. *Nature's colours*—use of natural colours and hues that represents natural contexts such as blues associated with clear skies and water, light browns associated with trees, grasses, savanna landscapes, greens associated with vascular plants that suggest flowering or fruiting bodies (Kellert, 2018).
6. *Information rich complexity*—inclusion of complex details representing the information richness of a natural environment setting, inclusive of organised complexity in geometric forms, shapes and nature representations (Fig. 4.3).

## Pattern Applications

There are various possibilities to integrate the biophilic attributes of the pattern Indirect Experience of Nature [2] into the urban environment (Table 4.1). The use of *Botanical Motifs* can be integrated in façade mouldings, column capitals, door and window frames, and street furniture. Use of *Biomorphic Ornament* in screens, lamp covers and streetlights, architectural screens and millwork. Include *Natural Geometry and Patterns* in walkway finishes, material patterns on façades and structures, and building typology. Representative of local geology, stone, soils and timber in the *Material and Texture* of walkways, retaining structures, steps, building structures, columns, entrances and interiors. Use of *Nature's Colours* such as shades of greens, light browns and soft blues throughout the cityscapes and building complexes. Avoid a monotone, dull and clinic exterior environment of the city by including *Information Rich Complexity* in building structures, façades, doors, windows, roof canopies, walkways and external furniture (Fig. 4.4).

**Fig. 4.3** Natural materials and texture. (Image author: PB Roös, 2022)

**Table 4.1**    Pattern comparison/relationship with others: INDIRECT EXPERIENCE OF NATURE [2]

| Roös (2022) | Indirect Experience of Nature [2][a] |
|---|---|
| Mehaffy et al. (2020) | Biophilic Urbanism [2.4] |
| | Local Symmetry [11.1] |
| | Fractal Pattern [11.3] |
| Salingaros (2019) | Fractals, Curves and Detail |
| | Representations-of-nature |
| | Organised-complexity |
| Kellert et al. (2008), Kellert (2018) | Indirect Experience of Nature |
| | Natural Patterns and Processes |
| Terrapin (Browning et al., 2014) | Biomorphic Forms and Patterns [8] |
| | Material Connection with Nature [9] |
| | Complexity and Order [10] |
| Alexander et al. (1977) | Good Materials [207] |
| | Floor-Ceiling Vaults [219] |
| | Roof Vaults [220] |
| | Column Place [226] |
| | Column Connections [227] |
| | Ornament [249] |
| | Warm Colours [250] |

[a] [2] indicates a *pattern*. When not indicated with [ ] the description in the table refers to an *attribute* or a *metric*

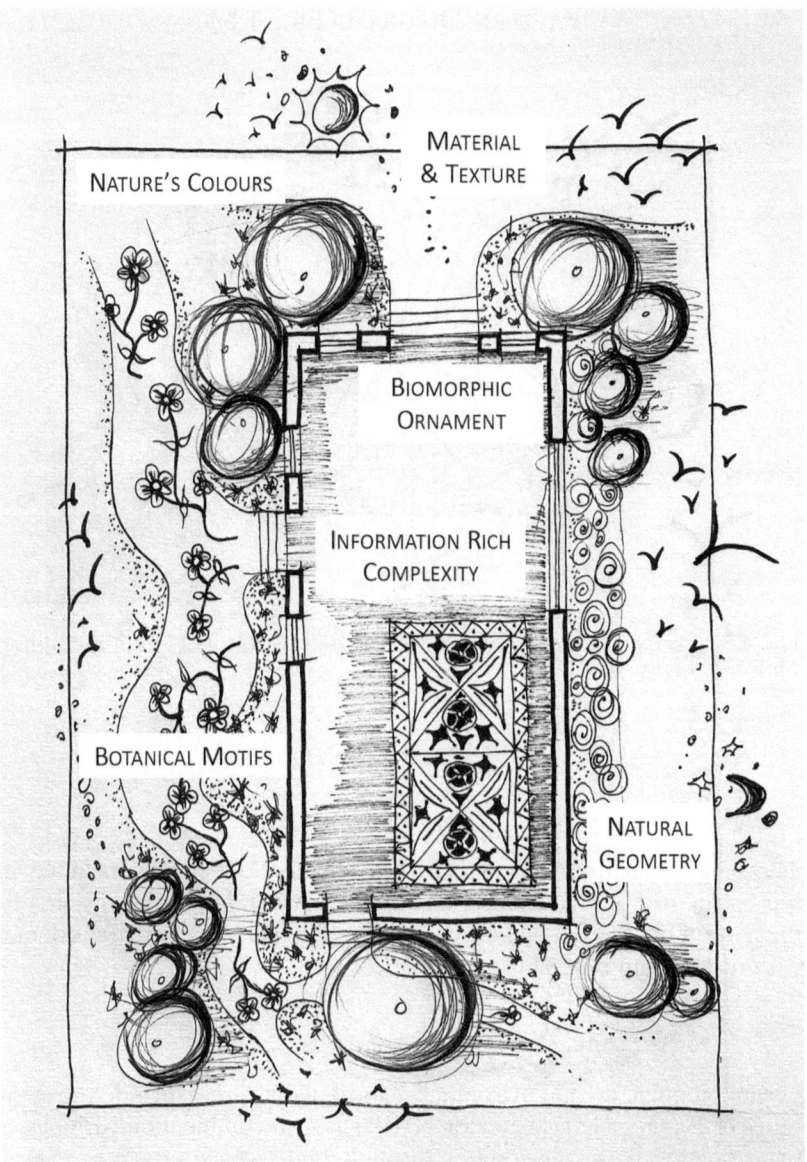

**Fig. 4.4** Application diagram of biophilic attributes for INDIRECT EXPERIENCE OF NATURE [2]. (Image author: PB Roös, 2022)

## PATTERN DIAGRAM (FIG. 4.5)

**Fig. 4.5**   Sketch of pattern—INDIRECT EXPERIENCE OF NATURE [2]. (Image author: PB Roös, 2022)

## PATTERN LINKS

### *Upward Links*

The biophilic pattern representative of elements of nature included in built structures—INDIRECT EXPERIENCE OF NATURE [2], links upwards to the pattern DIRECT EXPERIENCE OF NATURE [1], that involves actual contact with real attributes of nature.

### *Downward Links*

Create biophilic urban environments and buildings that include the attributes of INDIRECT EXPERIENCE OF NATURE [2], in combination with places that result in a sense of belonging through the EXPERIENCE OF PLACE, SPACE AND ATTACHMENT [3]. This pattern draws on the human ability to reason and attach meaning to the experience of a setting.

### *Integral Framework Alignment*

Most of the attributes of the pattern INDIRECT EXPERIENCE OF NATURE [2] align with the Integral Framework's upper right quadrant: *Nature Elements—Biologic*. This perspective includes the shapes, forms, geometries and elements in nature, referred to as the Biologic perspective. In the language of integral theory the 'It' is the actual representation of the elements in nature itself. However, our experiences from the connections to these elements can be aligned with the upper left quadrant: *Nature Experiences—Biophilia*, the individual person's direct experience referred to as the 'I'.

### *Comparison/Relationship with Patterns and Metrics by Other Authors*

Various other patterns can be aligned with the pattern INDIRECT EXPERIENCE OF NATURE [2]. For example, the pattern *Indirect Experience of Nature* by Kellert (2018) describes the need to include various representations from nature such as images, materials, texture, colour, shapes and forms, and elements that have weathered over time. Information richness—provides for information perceived from the richness of nature's most distinguishing characteristics (Kellert, 2018, p. 73). Biomorphic Forms and Patterns [8]—describes the symbolic references to contoured, patterned, textured or numerical arrangements that persist in nature; Material Connection with Nature [9]—provides for materials and elements from nature that through a minimal processing reflects the local ecology and geology to create a distinct connection to place; Complexity and Order [10]—identifies the rich sensory information that adheres to a spatial hierarchy similar to that encountered in wild nature (Browning et al., 2014). Table 4.1 attempts to align the pattern INDIRECT EXPERIENCE OF NATURE [2] with these other similar patterns and metrics, to assist the designer in the application of these in the city complexes.

## CONCLUSION STATEMENT

Humans are inherently part of the natural world, even though in a modern industrial society ignorant of nature, with a focus on technology and manufacturing, this innate human-nature connection continues to shape our evolution as a species and undeniably exists. It is this deeper connection to the organised complex order of abiotic and biotic systems that results in us *being nature*, or being part of the whole (Roös, 2021, p. 82). The result of this connection is the psychological phenomenon of a natural attraction, a sensory language that is formulated in a pattern language that enhances our wellbeing. Our biophilic design agenda thus needs to include the characteristics of a complex order and the complexity of the traditional ornament make-up (Salingaros, 2015), inclusive of the rules of living structures (Alexander, 2001–2005). Rules for how ornament make-up contributes to a healing environment can be derived from understanding how our brain is wired to respond to our surroundings (Roös, 2021, p. 84).

As noted by Beatley (2011), environments rich with nature and natural experiences (direct and indirect experience of nature) in the city will not only enhance wellbeing but will help strengthen commitments to sustainability and environmental awareness of our society (Beatley, 2011, p. 9). It is evident that if we want to create a sustainable future for our cities and the communities within them, we need to start at what enhances our own health and wellbeing—the DIRECT EXPERIENCE OF NATURE [1] and the INDIRECT EXPERIENCE OF NATURE [2].

*Therefore:*

*Incorporate biophilic elements that unconsciously trigger an emotional affiliation with the natural world through the representation of natural shapes, forms, pattern and processes in the city complexes. Use botanical motifs, biomorphic ornament, natural geometry and patterns, materials and texture, colour and information rich complexity to achieve the biophilic effect.*

## REFERENCES

Alexander, C. (2001–2005). *The nature of order—An essay on the art of building and the nature of the universe, Book One: The phenomenon of life.* Berkeley, California. USA: The Center for Environmental Structure.

Alexander, C., Ishikawa, S., Silverstein, M., Jacobson, M., Fiksdahl-King, I., & Angel, S. (1977). *A pattern language: Towns, buildings, construction.* Oxford University Press.

Beatley, T. (2011). *Biophilic cities—Integrating nature into urban design and planning.* Island Press.

Brethour, C., Watson, G., Sparling, B., Bucknell, D., & Moore, T. (2007). *Literature review of documented health and environmental benefits derived from ornamental horticulture products.* George Morris Centre.

Browning, W. D., Ryan, C. O., & Clancy, J. O. (2014). *14 Patterns of biophilic design*. Terrapin Bright Green, LLC.

Cohen, M. J. (2008). *Educating, counseling and healing with nature*. Project Nature Connect, Institute of Global Education.

Dravigne, A., Waliczek, T. M., Lineberger, R. D., & Zajicek, J. M. (2008). The effect of live plants and window views of green spaces on employee perceptions of job satisfaction. *HortScience (American Society for Horticultural Science)*, *43*(1), 183–187. https://doi.org/10.21273/HORTSCI.43.1.183

Kellert, S. R. (2018). *Nature by design*. Yale University Press.

Kellert, S. R., Heerwagen, J. H., & Mador, M. L. (Eds.). (2008). *Biophilic design. The theory, science and practice of bringing buildings to life*. Wiley.

Mattson, M. P. (2014). Superior pattern processing is the essence of the evolved human brain. *Frontiers in Neuroscience*, *8*, 265. https://doi.org/10.3389/fnins.2014.00265

Mehaffy, M. W., Salingaros, N. A., Kryazheva, Y., & Rudd, A. (2020). *A new pattern language for growing regions*. Sustasis Press.

Murchie, G. (1978). *Seven mysteries of life*. Houghton Mifflin.

Roös, P. B. (2021). *Regenerative-adaptive design for sustainable development—A pattern language approach* (Sustainable development goals series). Springer International. https://doi.org/10.1007/978-3-030-53234-5_1

Salingaros, N. A. (2015). *Biophilia and healing environments*. New York, NY: Terrapin Bright Green LLC and Amherst, MA: Levellers Press. https://www.terrapinbrightgreen.com/wp-ontent/uploads/2015/10/Biophilia-Healing-Environments-Salingaros-p.pdf

Salingaros, N. A. (2019). The biophilic index predicts healing effects of the built environment. *Journal of Biourbanism*, *8*(1), 13–34.

Sweeny, T. D., Grabowecky, M., & Suzuki, S. (2011). Awareness becomes necessary between adaptive pattern coding of open and closed curvatures. *Psychological Science*, *22*, 943–950. https://doi.org/10.1177/0956797611413292

Wang, W., Staffaroni, L., Reid, E., Jr., Steinschneider, M., & Sussman, E. (2009). Effects of musical training on sound pattern processing in high-school students. *International Journal of Pediatric Otorhinolaryngology*, *73*, 751–755. https://doi.org/10.1016/j.ijporl.2009.02.003

# CHAPTER 5

# Experience of Place, Space and Attachment [Pattern 3]

**Abstract** The experience of space and place attachment is one of the distinct human behaviours and is multi-dimensional, going beyond the visual perspective to include a multi-sensory experience. It is not simply a response to physical factors and perception, but depends on a complex and reciprocal relationship between experiences and behaviour, including cultural contexts. In cities it is possible to lose this 'deep attachment' due to a lack of exposure to elements of the natural environment. Place attachment arises as the result of *logical structures* from the cultivation of the meaning of environmental, architectural and cultural artefacts. This chapter designates the third meta biophilic pattern—EXPERIENCE OF PLACE, SPACE AND ATTACHMENT [3]—to guide city design to include logical structures and Indigenous Knowledge as part of biophilic design.

**Keywords** Placemaking • Sense of place • Topophilia • Biophilia • Indigenous knowledge • Attachment and belonging

© The Author(s), under exclusive license to Springer Nature Switzerland AG 2022
P. B. Roös, *A Biophilic Pattern Language for Cities*, Sustainable Urban Futures, https://doi.org/10.1007/978-3-031-19071-1_5

**Fig. 5.1**  Cultural connections to country. (Image author: PB Roös, 2022)

## INTRODUCTION

An individual's connection to place can be influenced by its surroundings, environmental conditions, and both social and cultural factors (Fig. 5.1). Leyden et al. (2011) highlighted the importance of community recreational spaces where individuals can congregate and interact with each other. Places that promote socialising enable people to feel connected to, secure, and trusting of one another. Further, when places of community gatherings (and spaces) include acknowledgement of the local Indigenous peoples, the initial Traditional Owners of the land and a deeper connection to place—a process of healing occurs. Engaging with and feeling connected to one's place as well as a sense of community have been found to contribute to individual wellbeing (Bornioli et al., 2018; Leyden et al., 2011). In addition, an attractive physical environment that includes an *abundance of natural features* for leisure activities has also been found to improve community health and wellbeing (Bornioli et al., 2018). Places with deep attraction result in a place that has 'a sense of place'.

When people refer to a place that has 'a sense of place', they are referring to an emotional attachment they feel for that specific setting. Other references to place attachment include when a place has 'a lot of character', 'a place has soul' and 'a place with feeling'. The degree to which a place has these characteristics and the effect on a person is mostly associated with how distinctive its environmental, cultural and social features are (Roös, 2021; Green, 2010). This element of 'character' is evident in all natural and built environments, even places that evoke a depressing or negative feeling—a place with 'no character'. Places considered to have recognisable qualities and a unique character, as noted above, are referred to as having a 'sense of place' (Roös, 2021, p. 49).

These places evoke a strong attachment, and many of these places even link to spiritual and religious contexts. 'Sense of place' is a term developed from the term *genius loci,* a term initially used to describe the appreciation of landscapes (Jiven & Larkham, 2003). *Genius loci* can be acknowledged as a technical term, specific to the natural landscape. However, because this term developed over time, the concept moved away from only 'a natural landscape', and included its application to any landscapes, including the urban form (Jackson, 1994). *Genius loci,* in its application to perception of a place, evolved into describing the *quality of places* and in the terminology's transition to modern times it evolved into 'sense of place' (Roös, 2021).

The pattern EXPERIENCE OF PLACE, SPACE AND ATTACHMENT [3], represents these experiences of attachment, places that have unique character, places of meaning, places that have soul, places that are sacred. Embedded in these places is the quality to provide healing and assist in wellbeing, places that attract and demand a place of return, as noted by Jackson (1994):

'Sense of Place' is a much used expression, chiefly by architects but taken over by urban planners and interior decorators and the promoters of condominiums, so that now it means very little. It is an awkward and ambiguous translation of the Latin term *genius loci*. In classical times it means not so much the place itself as the guardian divinity of that place. ... in the eighteenth century the Latin phrase was usually translated as 'the genius of a place', meaning its influence. ... We now use the current version to describe the atmosphere to a place, the quality of its environment. Nevertheless, we recognize that certain localities have *an attraction which gives us a certain indefinable sense of well-being and which we want to return to, time and again* (Jackson, 1994, p. 64).

## PATTERN STATEMENT

*To be attached and belong requires spaces that demand a sense of attraction and return, places that provide a certain indefinable sense of wellbeing and attachment, a sense of deep connection to the environment, and places that celebrate local culture and the Indigenous peoples of the land.*

## DISCUSSION

Every day in our cities millions of people travel through cityscapes of concrete, glass and neon mazes that are embellished with adverts and brands that create an image of a world of consumption. Just as these brands and products are based somewhere else, detached from the place where they are displayed and used, connection to locality in an industrialised world is nearly impossible (Sampson, 2012). Globalisation and rapid urbanisation have normalised the industrial world, and as a result support this detached condition. The current situation is a loss of contact with nature, personal and cultural identity. This condition of urbanites is what Pyle (1998) has dubbed "the extinction of experience". Engulfed in a sea of monotonous built environments and sameness, humans lack attachment to places in which they live and work (Sampson, 2012, p. 23). Many city environments can be qualified as places with no life. However, various places in cities do have a sense of character, mostly the places that are historical and include old traditional buildings, gardens, public parks and gathering spaces.

We know that in some places that we visit we experience a good feeling from the place, and in others we don't experience positive feelings. Some authors propose that it is the visual stimulation that predominantly causes this experience of good feelings and argued this phenomenon as a perception of 'beauty' (Ruggles, 2019). The concept of beauty in architecture is highly debatable, and after a long pause in practice and literature, the subject and questions of beauty are surfacing again in the current climate of uncertainty, and obvious unhealthy built environments. The simplistic buildings with extensive smooth glass facades, concrete and steel structures, minimalistic surfaces and other conspicuous consequences of prevailing modernist design ideals have led to confusion about 'beauty' and a connection to place (Salingaros, 2020). It is not surprising that these 'simplistic' typologies embedded in abstraction, formalism and modernism with a focus on surface appearance result in a 'fight or flight' reaction, causing stress in our physical and psychological state (Salingaros, 2020; De Paiva & Jedon, 2019).

However, deep emotional experience of place goes beyond the visual connection of that place, it includes the notion of inner feeling, which we connect to unconsciously (Alexander, 2001–2005). Christopher Alexander argued that the principles of *'living centres'* or *'living structures'* of a place, which connect human built environments and nature together in a collective wholeness, add to the creation of place character and a sense of belonging (Roös, 2021; Alexander, 2001–2005, p. 80). The living structures within an urban setting are crafted over many years by residents, resulting in a unique character embedded in the heritage and history of the place. This uniqueness is fundamental to the ongoing resilience of a place and results in the *sustainable futures of the communities* (Alexander, 2004). An important aspect of place-character is the heritage and cultural identity of place, shaped by the history of the settlement.

Throughout history most of humanities occupancy has occurred in the habitats of First Nation peoples who were directly tied to their natural environment. In addition, Indigenous communities were living in harmony with the land and place, having deep knowledge of the natural environment, seasons, climate, and long-term cycles of evolution. In many instances in current times this is still the case, and Indigenous knowledge systems are used for the detailed understanding of place (Sampson, 2012; Settee, 2008). This deep connection to place can be named as *'Topophilia'* meaning an emotional bond with place, and as defined by Yi-Fu Tuan (1974 [1990]) as "the effective bond between people and place or setting" (p. 4). According to Alexander (1964), emotional connection to place involves the creation of *logical structures*, as has been used by Indigenous cultures (Alexander, 1964). Indigenous cultures repeat patterns of tradition, where form is informed by symbols, and the system passes through history adapting forms to the context of the environment

(Van der Ryn, 2005). A good example of these complex patterns and forms appears continuously in representations of complex fractals in settlement architecture, cross-cultural engagements and logarithmic scaling in designs in African cultures (Eglash, 1999). The creation of logical structures in our urban environments can then result in the phenomena of *topophilia*—a deep emotional attachment to place supported by biophilic attributes and shapes.

Design patterns, that include various biophilic attributes can be used to enhance a person's EXPERIENCE OF PLACE, SPACE AND ATTACHMENT [3], stimulating deep emotional experiences with place (*topophilia*), including social and cultural connections. In *A Pattern Language* (Alexander et al., 1977), Alexander and his colleagues provide various design patterns that promote cultural and community engagement such as Small Public Squares [61], Birth Places [65], Holy Ground [66] and Common Land [67]. Biophilic attributes can include elements of design that promote connection to place including in a social and cultural context (Fig. 5.2).

## BIOPHILIC ATTRIBUTES

The biophilic attributes resulting in an EXPERIENCE OF PLACE, SPACE AND ATTACHMENT [3] can include those elements that unconsciously trigger a sense of attachment and belonging to a specific place setting. In this instance it is not only the biophilic attributes that result in the experience, but a combination of the biophilic attributes with social and cultural placemaking attributes that result in a sense of place (and belonging). The result is a design pattern (rather than a biophilic pattern) that informs placemaking supported by biophilic attributes as follow:

1. *Celebration of Indigenous Knowledge*—inclusion of local Indigenous Knowledge in designing places with a deeper reading of the natural environment, pre-European settlement timeline, pedagogical landscape attributes and reference to *Country*.[1]

---

[1] *Country*—is a term used by the Aboriginal people of Australia to refer to the land to which they belong and the place of their dreaming (Smyth, 2004).

**Fig. 5.2**  Symbolic reference. (Image author: PB Roös, 2022)

2. *Historical and Cultural Artifacts*—provision of various artefacts, mouldings, materials or artworks that represents the history and cultural heritage of the place, including local content. Include biophilic attributes of biomorphic forms, fractal geometry, ornament and symbolic objects.

3. *Gathering Circles*—meeting places for knowledge sharing and storytelling, gathering circles[2] within the urban landscape, gardens, parks and courtyards that represent the gathering of people around a firepit or similar central object with circular seating made out of natural materials. Include biophilic attributes of natural materials, native vegetation, weather patterns, access to sunlight and a direct line of sight to the sky.

4. *Gathering Ground*—places of common community areas for small gatherings and social activities between buildings, in courtyards, on roof tops, and in alcoves. Include biophilic attributes of natural materials, native vegetation, water features and access to sunlight (Fig. 5.3).

## PATTERN APPLICATIONS

There are various possibilities to integrate the biophilic attributes of the pattern EXPERIENCE OF PLACE, SPACE AND ATTACHMENT [3] into the urban environment (Fig. 5.4). *Celebrate Indigenous Knowledge* by firstly identifying and engaging with the local Indigenous people of the land or place you are on, discussing and identifying the elements to reference *Country*, the opportunities for a deep connection to *Country*, as well as how the landscape/cityscape can be a pedagogical space to learn of *Country*. Include *Historical and Cultural Artifacts* to connect place to history, to

---

[2] *Gathering Circles*—is based on the traditional *Yarning Circles* used by Aboriginal and Torres Strait Islander peoples for thousands of years as a conversational process that involves the telling of stories as a way of passing on cultural knowledge. These circles provide a safe place for all to speak without judgement (University of Newcastle, 2022).

**Fig. 5.3** Gathering place—Wadawurrung Country. (Image author: PB Roös, 2022)

support 'truth telling', create meeting spaces that include *Gathering Circles* for storytelling and small community *Gathering Grounds* throughout the city environments. These spaces need to be complimented with biophilic attributes that enhance the place settings resulting in a sense of wellbeing.

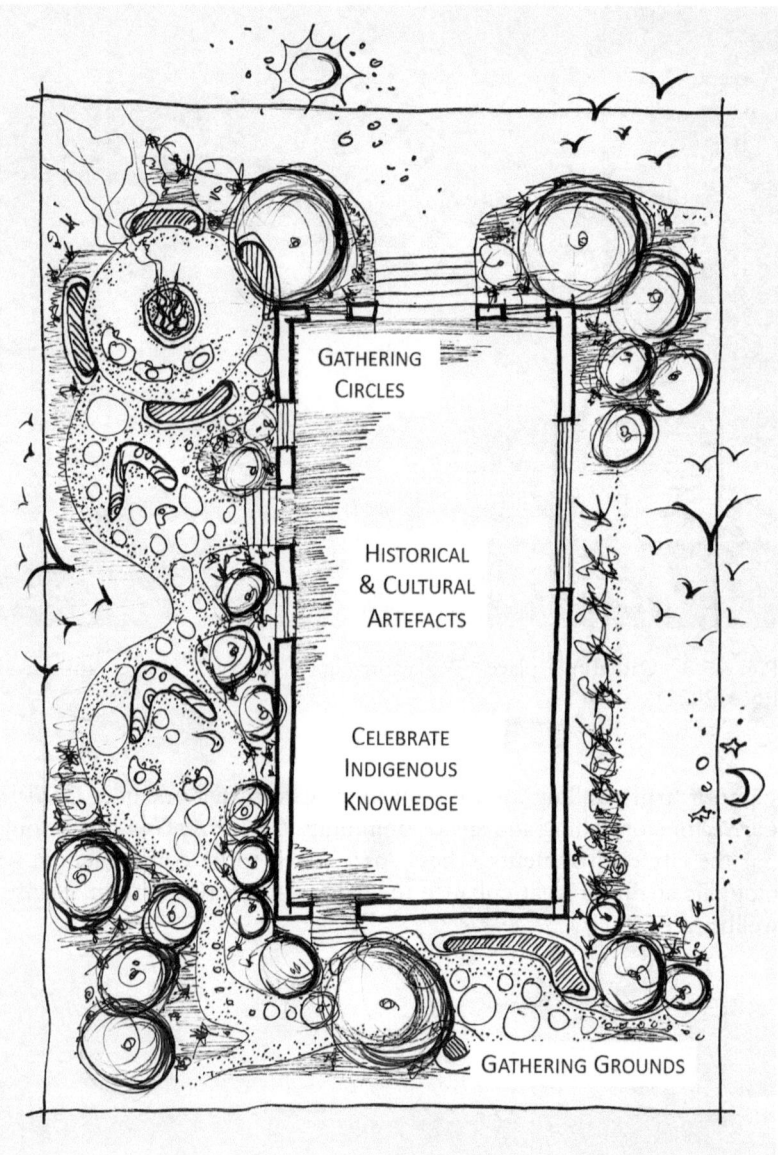

**Fig. 5.4** Application diagram of biophilic attributes for EXPERIENCE OF PLACE, SPACE AND ATTACHMENT [3]. (Image author: PB Roös, 2022)

## Pattern Diagram (Fig. 5.5)

**Fig. 5.5** Sketch of pattern—Experience of Place, Space and Attachment [3]. (Image author: PB Roös, 2022)

## Pattern Links

### Upward Links

The biophilic pattern representative of historical and cultural emotional attachment to place—Experience of Place, Space and Attachment [3], links upwards to the pattern Indirect Experience of Nature [2], that incorporates biophilic elements that unconsciously trigger an emotional affiliation with the natural world.

### Downward Links

Create spaces that include the attributes of Experience of Place, Space and Attachment [3], in combination with design elements that support ecological sustainability and planetary wellbeing through the application of the pattern NATURE PATTERNS, PROCESSES AND SYSTEMS [4]. This pattern fosters a collective of nature's contexts in a larger ecological system to sustain life.

### Integral Framework Alignment

The attributes of the pattern EXPERIENCE OF PLACE, SPACE AND ATTACHMENT [3], aligns with the Integral Framework's lower left quadrant: *Nature Meanings—Bioculture.* This perspective includes the meaning of connection to place through the lens of Indigenous Knowledge and systems of the land, referred to as the Bioculture perspective. In the language of integral theory it is the 'We', the meaning of the connection to the natural world, connection to *Country,* the place of belonging and attachment, the 'reason' for placemaking.

### Comparison/Relationship with Patterns and Metrics by Other Authors

The perspective of *deep connection to place* and the consideration of Indigenous Knowledge in the context of biophilic design—is absent in the literature. In fact as far as I am aware, there is no relationship drawn between the deep human-nature connection in the biophilia hypothesis, and the deep connection to nature by Indigenous cultures, where people and *Country* are one and the same, by other authors. If biophilic design is an attempt to reconnect people with nature, then the first step should be to acknowledge the knowledge of place held by our Indigenous communities. If you cannot establish a real, deep cultural and emotional connection with nature and place, your biophilic design attempt is doomed to fail and will be superficial. It is therefore important to apply the key principles of the pattern EXPERIENCE OF PLACE, SPACE AND ATTACHMENT [3], recognising the other supportive patterns and attributes that can achieve a strong sense of attachment and belonging to place. The following patterns and attributes listed by others (Table 5.1), should therefore be read with the understanding that these are only supportive or complimentary, and to be used to help advance the real attachment to place—*topophilia*—a deep emotional cultural connection. Table 5.1 attempts to align the pattern EXPERIENCE OF PLACE, SPACE AND ATTACHMENT [3], with other similar patterns and metrics, to assist the designer in the application of these.

**Table 5.1**  Pattern comparison/relationship with others: EXPERIENCE OF PLACE, SPACE AND ATTACHMENT [3]

| Roös (2022) | Experience of Place, Space and Attachment [3][a] |
|---|---|
| Mehaffy et al. (2020) | Malleability [12.4] |
| | Informal Stewardship [18.4] |
| Salingaros (2019) | Organised-complexity |
| Kellert (2018) and Kellert et al. (2008) | Experience of Place and Space |
| | Place-based Relationships |
| Terrapin (Browning et al., 2014) | Complexity and Order [10] |
| | Prospect [11] |
| | Refuge [12] |
| | Mystery [13] |
| | Awe [15] |
| Alexander et al. (1977) | Small Public Squares [61] |
| | Birth Places [65] |
| | Holy Ground [66] |
| | Common Land [67] |

[a] [3] indicates a *pattern*. When not indicated with [ ] the description in the table refers to an *attribute* or a *metric*

## CONCLUSION STATEMENT

The experience of space and place attachment is one of the distinct human behaviours and is multi-dimensional, going beyond the visual perspective to include a multi-sensory experience. It is not simply a response to physical factors and perception but depends on a complex and reciprocal relationship between experiences and behaviour, including cultural contexts (Roös, 2021; Lewicka, 2011). Further influences on attachment to place come from the direct surrounding environment, including elements of nature and biophilia. Sense of place attachment arises as the result of cultivation of the meaning and the environmental and architectural artefacts, associated with created places that include cultural values (Giuliani, 2016). The EXPERIENCE OF PLACE, SPACE AND ATTACHMENT [3], includes emotional attachment achieved through the creation of *logical structures* as part of *topophilia* and supportive biophilic attributes.

*Therefore:*

*To create places of meaning requires in the first instance acknowledgement of the Indigenous people of the land and the identification of the elements to reference Country. Use design patterns and biophilic attributes to celebrate place using local historical elements including artifacts, symbols, materials and environmental features to allow for a deeper reading of the natural environment, pre-European settlement timeline, pedagogical landscape attributes, and reference to Country.*

## REFERENCES

Alexander, C. (1964). *Notes on the synthesis of form.* Harverd University Press.
Alexander, C. (2001–2005). *The nature of order—An essay on the art of building and the nature of the universe, Book One: The phenomenon of life.* The Center for Environmental Structure.
Alexander, C. (2004). *Sustainability and morphogenesis: The birth of a living world.* Centre for Environmental Structure.
Alexander, C., Ishikawa, S., Silverstein, M., Jacobson, M., Fiksdahl-King, I., & Angel, S. (1977). *A pattern language: Towns, buildings, construction.* Oxford University Press.
Bornioli, A., Parkhurst, G., & Morgan, P. L. (2018). The psychological wellbeing benefits of place engagement during walking in urban environments: A qualitative photo-elicitation study. *Health and Place, 53,* 228–236.
Browning, W. D., Ryan, C. O., & Clancy, J. O. (2014). *14 Patterns of biophilic design.* Terrapin Bright Green, LLC.
De Paiva, A., & Jedon, R. (2019). Short- and long-term effects of architecture on the brain: Toward theoretical formalization. *Frontiers of Architectural Research, 8*(4), 564–571. https://doi.org/10.1016/j.foar.2019.07.004
Eglash, R. (1999). *African fractals: Modern computing and indigenous design.* Rutgers University Press.
Giuliani, M. V. (2016). *Psychological theories for environmental issues* (pp. 137–169). Ashgate Publishing.
Green, R. J. (2010). *Coastal towns in transition—Local perceptions of landscape change.* CSIRO Publishing.
Jackson, J. (1994). *A sense of place, a sense of time.* Yale University Press.
Jiven, G., & Larkham, P. J. (2003). Sense of place, authenticity and character: A commentary. *Journal of Urban Design, 8*(1), 68–71.
Kellert, S. R. (2018). *Nature by design.* Yale University Press.
Kellert, S. R., Heerwagen, J. H., & Mador, M. L. (Eds.). (2008). *Biophilic design. The theory, science and practice of bringing buildings to life.* Wiley.
Lewicka, M. (2011). Place attachment: How far have we come in the last 40 years? *Journal of Environmental Psychology, 31*(3), 207–230.
Leyden, K. M., Goldberg, A., & Duval, R. D. (2011). The built environment, maintenance of the public sphere and connections to others and to place: An examination of 10 international cities. *Journal of Urbanism, 4*(1), 25–38.

Mehaffy, M. W., Salingaros, N. A., Kryazheva, Y., & Rudd, A. (2020). *A new pattern language for growing regions*. Sustasis Press.

Pyle, R. M. (1998). *The thunder tree: Lessons from an urban wildland*. Lyons Press.

Roös, P. B. (2021). *Regenerative-adaptive design for sustainable development—A pattern language approach* (Sustainable development goals series). Springer International. https://doi.org/10.1007/978-3-030-53234-5_1

Ruggles, D. H. (2019). Beauty, neuroscience and architecture. In I. Palti (Ed.), *The conscious cities anthology*. The Centre for Conscious Design. https://doi.org/10.33797/CCA19.01.08

Salingaros, N. A. (2019). The biophilic index predicts healing effects of the built enviroment. *Journal of Biourbanism, 8*(1), 13–34.

Salingaros, N. A. (2020). Connecting to the world: Christopher Alexander's tool for human-centered design. *She Ji: The Journal of Design, Economics, and Innovation, 4*, 455–480.

Sampson, S. D. (2012). The topophilia hypothesis: Ecopsychology meets evolutionary psychology. In P. H. Kahn Jr. & P. H. Hasbach (Eds.), *Ecopsychology: Science, totems and the technological species* (pp. 23–53). MIT Press.

Settee, P. (2008). Indigenous knowledge as the basis for our future. In M. K. Nelson (Ed.), *Original instructions: Indigenous teachings for a sustainable future* (pp. 42–47). Bear.

Smyth, D. (2004). *Kooyang sea country plan*. Framlingham Aboriginal Trust and Winda Mara Aboriginal Corporation.

Tuan, Y. F. (1990). *Topophilia: A study of environmental perceptions, attitudes, and values*. Columbia University Press. (Original work published 1974).

University of Newcastle. (2022). *Yarning circle*. Retrieved August 12, 2022 from https://www.newcastle.edu.au/campus-life/central-coast/ourimbah/spaces-and-places/yarning-circle

Van der Ryn, S. (2005). *Design for life*. Gibbs Smith.

CHAPTER 6

# Nature Patterns, Processes and Systems [Pattern 4]

**Abstract** A deep affiliation and connection to place includes two dimensions: the biophilia dimension and the topophilia dimension. As a result these dimensions demand a relationship with different levels of scale in place attachment, from the local context extending to the features of a larger ecological system, including prominent landmarks, landscapes, natural features, geological forms, mountains and rivers, climate cycles, seasons, prevailing weather patterns, local fauna and flora and place-based natural systems. When designing places in the city consideration for the larger interconnected network of processes and ecological systems needs to be applied. This chapter designates the fourth meta biophilic pattern— NATURE PATTERNS, PROCESSES AND SYSTEMS [4]—to guide city design to include a regenerative response to placemaking, where design considers the community positioned in a larger ecological scale.

**Keywords** Ecology • Ecological design • Topophilia • Biophilia • Levels of scale • Ecological systems

P. B. Roös, *A Biophilic Pattern Language for Cities*, Sustainable
Urban Futures, https://doi.org/10.1007/978-3-031-19071-1_6

**Fig. 6.1**    Nature's patterns. (Image author: PB Roös, 2022)

## Introduction

A person's emotional attachment to place typically includes two dimensions that can be defined as the *biophilia* and *topophilia* dimensions (Fig. 6.1). The biophilia dimension supports connection to place that can be stressed as natural, geographical and physical. The topophilia dimension supports connection to place that can be stressed as the cultural, social and historical affinities, amongst others (Kellert, 2018; Sampson, 2012). However, topophilia also links strongly to geographical locations, and combined with biophilia—the innate human affiliation with nature—attachment becomes a much more complex phenomena where the experience of deep connection to place extend to the features of a larger ecological system, including prominent landmarks, landscapes, natural features, geological forms, mountains and rivers, climate cycles, seasons, prevailing weather patterns, local fauna and flora, and place-based natural systems.

Considering that the extended network of connection with ecological systems is as a result of biophilia and topophilia, biophilic design is not only about enhancing *human* health and wellbeing, but also directly linked to the health and wellbeing of the *more-than-human*—in essence all beings as part of nature, and the larger earth system that sustains life. When designing places for habitation, we need to consider the larger interconnected network—the *whole* system. From an ecological perspective, these connections of the larger framework (scale linking) of 'the whole' are dealt with in levels of scale at the larger ecosystem, and at a certain local place-based scale at the location of place (Lyle, 1991 [1985]). To work within the larger context, a set of hierarchical scales in a larger organised system is applied. As part of biophilic design thinking, the context of *scale* and *order* are fundamental in the process of design, which considers *hierarchies of scale* and represents in the midst of complexity, the nature of *order* (Roös, 2021, p. 103). This phenomenon of order as suggested by Christopher Alexander (2001–2005) propositions a pattern language, and this language connects the parts to create wholes. This biophilic concept of 'connecting parts to wholes', as stated by Kellert (2018) is a process part of a larger system in creating connections of 'place character' to comprise a larger ecosystem that results in an ecological experience (p. 105). The biophilic pattern NATURE PATTERNS, PROCESSES AND SYSTEMS [4] provides us guidance in implementing a regenerative response to placemaking where design considers the place (community) positioned in a larger landscape and ecological scale.

## PATTERN STATEMENT

*Deep connection to place includes biophilia and topophilia that extends to the larger ecological systems around us including prominent landmarks, landscapes, natural features, geological forms, mountains and rivers, climate cycles, seasons, prevailing weather patterns, local fauna and flora, and place-based natural systems.*

## Discussion

Considering the vast expanse of many cities, the interference in the ecology on their fringes and the regions they are located in undoubtably has a direct impact on the environment. The design and planning of cities needs to embrace an ecological design process to mitigate the impacts of the built environment on natural systems. Acknowledging this crisis in the wake of global urbanisation in the late 60s—Ian McHarg (1967) proposed an 'ecological approach' for urban and regional planning. The ecological approach methodology aimed to reveal 'nature as a process containing intrinsic forms' (Roös, 2021, p. 93). The method of McHarg involved a *set of layers* similar to layers that appear in nature, starting with the *geology of a place* and concluding with the *habitats of plants, animals and humans* (McHarg, 1992 [1969]). This ecological method offered opportunities and constraints having regard for a specific place of habitation and human interference, and that plants, animals and humans in the specific location and setting are only comprehensible in terms of their physical and biological evolution (McHarg, 1967).

As mentioned earlier in this chapter, when ecology is part of biophilic design thinking, the context of *scale* and *order* are fundamental in the process of placemaking specific to location, which considers *hierarchies of scale*, which in turn directly aligns with McHarg's set of layers that provides a *cross-ecological inventory* that can be interpreted as analysing existing and proposed human land use, design and habitation (Swaffield, 2002, p. 39; McHarg, 1967). This ecological inventory summarises the patterns of place, and enables planning for *creative fitting*. *Creative fitting* locates the fittest environment for both human and natural uses, integrated so that the uses of place fit in with a changed landscape over the scale of time. McHarg's 'rationale method' followed a sequence that includes 'layers' of place in an order of climate, geology, physiography, hydrology, soils, plant ecology, wildlife habitats and land use (human settlement). Analysis of place using this order allows for establishing a deep human connection to place, as we can note in his explanation of *creative fitting*:

> Written on the place and upon its inhabitants lies mute all physical, biological and cultural history to be understood by those who can read it. This is the prerequisite for intelligent intervention adaptation. So, let us begin at the beginning. The place, any place, can only be understood through its physical evolution. Both climate and geology can be invoked to interpret

physiography, the current configuration of the place. If one knows historical geology, climate, and physiography, then the water regimen becomes comprehensible: the pattern of rivers and aquifers, their physical properties and relative abundance, oscillation between flood and drought. Knowing the foregoing and the prior history of plant evolution, we can now comprehend the nature and pattern of soils. By identifying physiographic, climatic zones and soils, we can perceive order and predictability in the distribution of constituent plant communities. Animals are fundamentally related so that given the preceding information, with the addition of the stage of succession of the plant communities and their age, it is possible both to understand and to predict the species, abundance or scarcity of wild animal populations (McHarg, 1992 [1969], pp. 105–7).

Any newly built structure interferes with the ecology of a place, as a single structure or building and more so as a settlement or a large city. The pattern NATURE PATTERNS, PROCESSES AND SYSTEMS [4] brings together the criteria for analysis of place as per the description of McHarg's 'rationale method' above (Fig. 6.2).

## BIOPHILIC ATTRIBUTES

The biophilic attributes as part of the pattern NATURE PATTERNS, PROCESSES AND SYSTEMS [4] include those elements that connect people to places of geographical and ecological context, including relating the city to the larger regional ecological system. The result is patterns of ecological connection supported by biophilic attributes that inform placemaking as follows:

1. *Ecological landscape connections*—inclusion of ecosystems and landscape ecologies in designing places with a direct link to prominent ecosystems and biogeographical features such as watersheds, wetlands, mountains, rivers, estuaries and oceans. Include biophilic attributes with the presence of water, local and native flora and fauna, views of natural elements and processes including the exposure to weather.
2. *Geographical place connections*—provision of features in geological representations, exposure of geology, view lines to prominent landscape features, siting at prominent landscapes, orientation.

**Fig. 6.2**   Wetlands: ecological systems of place. (Image author: PB Roös, 2022)

3. *Local habitats and systems*—design places that contribute to the local biodiversity by including native plants that support local fauna and flora habitats, enhancing biological productivity, siting in appropriate distance and proximity to sensitive ecosystems, providing optimal scale and size, design for regeneration.

4. *Pattern Scale Linking*—places of appropriate scale within the larger functioning ecological system of place, hierarchical patterns connected at different levels and scales, including arithmetically or geometrically related. Complex patterns of locality can be expressed in the arithmetical expressions of local natural and built forms that include the golden proportion of the Fibonacci ratio (Kellert et al., 2008; Portoghesi, 2000). Additional biophilic patterns of biomorphic forms, fractal geometry, and complexity can enhance and support pattern scale linking (Fig. 6.3).

## PATTERN APPLICATIONS

There are various possibilities to integrate the elements and biophilic attributes of the pattern NATURE PATTERNS, PROCESSES AND SYSTEMS [4] into the urban environment, as well as the suburban and city fringes (Fig. 6.4). *Ecological Landscape Connections* assist in linking to the local ecosystems and biogeographical features, whereas *Geographical Place Connections* enhance place attachment by providing geological and prominent landscape feature connections. Include native plants and environmental considerations that contribute to the local biodiversity and habitats by design with *Local Habitats and Systems*. Consider the *Pattern Scale Linking* of all ecosystems, processes and patterns at different levels and scales. Use biophilic attributes that strengthen the complexity of the systems, including geometries that support different levels of scale.

**Fig. 6.3** Different habitats—Geelong Botanic Gardens. (Image author: PB Roös, 2022)

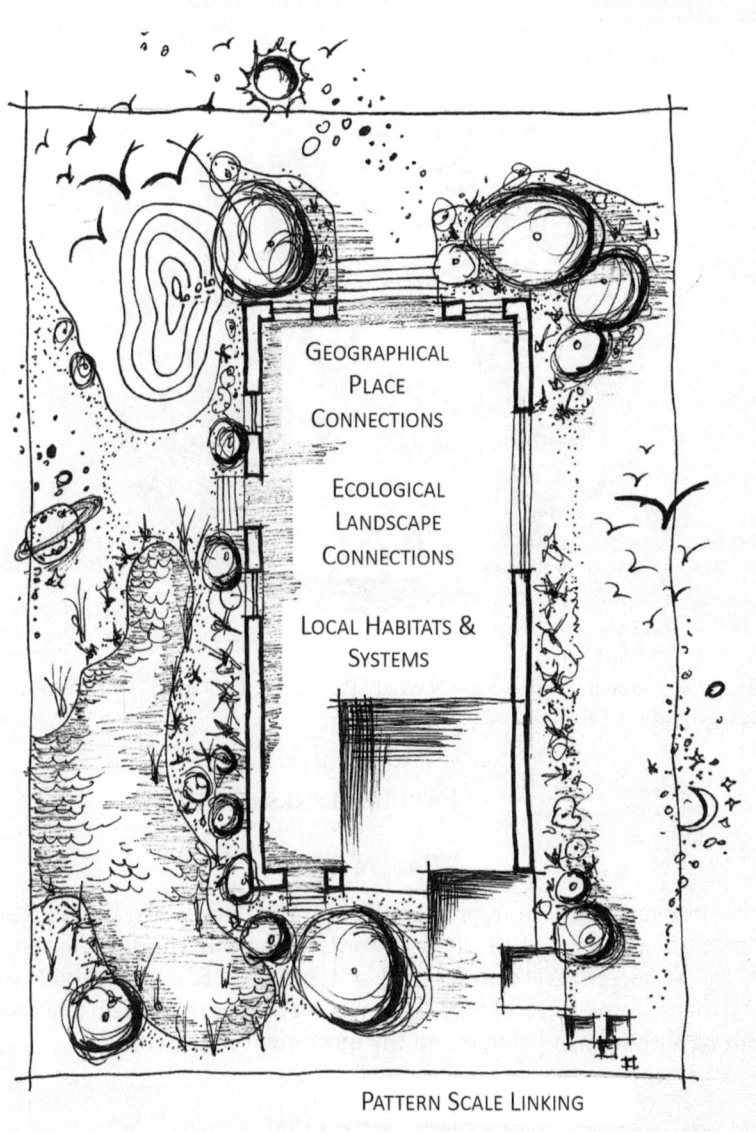

**Fig. 6.4**  Application diagram of biophilic attributes for—NATURE PATTERNS, PROCESSES AND SYSTEMS [4]. (Image author: PB Roös, 2022)

## Pattern Diagram (Fig. 6.5)

**Fig. 6.5** Sketch of pattern—Nature Patterns, Processes and Systems [4]. (Image author: PB Roös, 2022)

## Pattern Links

### *Upward Links*

The biophilic pattern representative of the local—Nature Patterns, Processes and Systems [4] links upwards to the patterns of Experience of Place, Space and Attachment [3] and Indirect Experience of Nature [2], that incorporate biophilic attributes that support a sustainable ecosystem of both human habitats and the more-than-human habitats.

### *Downward Links*

Nature Patterns, Processes and Systems [4], links with the first pattern Direct Experience of Nature [1] to complete the cycle of interconnectivity across all scales, dimensions and applications, in combination with design elements that support ecological sustainability and planetary wellbeing. This pattern fosters a collective of nature's contexts in a larger ecological system to sustain life.

### Integral Framework Alignment

The attributes of the pattern NATURE PATTERNS, PROCESSES AND SYSTEMS [4], align with the Integral Framework's lower right quadrant: *Nature Contexts—Biosystem*. This perspective includes the understanding of the *regenerative process* of a *whole system* that is part of the larger ecological system of earth—the *Biosystem*. Patterns of nature stipulate a *place-based* understanding of the *integrated human and ecological systems*, which can only be appreciated through exploring the nature of regeneration processes and their relationship to patterns and a 'biophilic pattern language' of nature and the built environment.

### Comparison/Relationship with Patterns and Metrics by Other Authors

In Biophilic Cities (2011) Timothy Beatley dedicated a chapter to explaining the nature in cities (Beatley, 2011, pp. 17–43), and noted the presence of nature and how we as humans experience nature as the *nature above*, the *nature all around*, the *nature below* and the *nature behind (and ahead)*. On closer observation, it becomes clear that these different levels of experiencing nature in our cities are in fact aligned with the biophilic patterns and attributes that allow us to connect with the full context of the natural world not only in what we can see, but also in the workings of the ecological systems, processes and patterns of nature. The *Blue-Green Network* [1.2], and *Urban Greenway* [3.1] patterns by Mehaffy et al. (2020) link us to the *nature all around* and *nature below* in the context of vegetation, trees, biodiverse corridors and water bodies, and streams and rivers in the city, whereas the biophilic pattern *Connections with Natural Systems* [7], of Browning et al. (2014) link us to the *nature behind (and ahead)*. As noted by Beatley "every bit of nature visible in cities today, every landform or hydrologic feature, is to some degree shaped by the life history of that point on Earth... informed by a deeper, longer process of evolution and change and reformulation" (Beatley, 2011, p. 39).

**Table 6.1**   Pattern comparison/relationship with others: NATURE PATTERNS, PROCESSES AND SYSTEMS [4]

| Roös (2022) | Nature Patterns, Processes and Systems [4][a] |
| --- | --- |
| Mehaffy et al. (2020) | Blue-Green Network [1.2] |
| | Urban Greenway [3.1] |
| Salingaros (2019) | Life |
| Kellert (2018) and Kellert et al. (2008) | Integrating Parts to Create Wholes |
| | Landscapes |
| | Organised Complexity |
| Terrapin (Browning et al., 2014) | Presence of Water [5] |
| | Connections with Natural Systems [7] |
| Alexander et al. (1977) | City Country Fingers [3] |
| | Agricultural Valleys [4] |
| | Access to Water [25] |
| | Still Water [71] |
| | Animals [74] |
| | Terraced Slope [169] |
| | Tree Places [171] |
| | Garden Growing Wild [172] |
| | Garden Wall [173] |

[a] [4] indicates a *pattern*. When not indicated with [ ] the description in the table refers to an *attribute* or a *metric*

Similarly the biophilic attribute (or metric)—*Life*—noted by Nikos Salingaros (2019) can be considered to align with various patterns to represent the conditions of life-giving structures in a city, and can link us with all levels of nature in the city including the *nature above*, the *nature all around*, the *nature below* and the *nature behind (and ahead)*. Table 6.1 below attempts to align the pattern NATURE PATTERNS, PROCESSES AND SYSTEMS [4], with other similar patterns and metrics, to assist the designer in the application of these, in essence to connect deeply with the life-giving phenomena of Earth.

## CONCLUSION STATEMENT

In the previous chapter we have identified that sustainability needs to progress to a regenerative status. This was clearly articulated by McDonough and Braungart (2002), where they report that narratives about sustainability have the tendency to include the typical words of reduce, minimise, limit, avoid, less harm and sustain. The emphasis and language of today's sustainability thinking is largely one of reducing resource use and adverse environmental damage by minimising human industries and systems (Cole, 2012; McDonough & Braungart, 2002). In other words, the notion is to only increase efficiency to make systems less damaging. Sustainability is not likely to be achieved through a reduction of human impact alone (du Plessis & Cole, 2011; du Plessis, 2012). It is not enough, neither inspiring to focus on mitigating the effects of human activity alone, and it is still degenerating the natural environment (Roös & Jones, 2013).

From this perspective, it is critical to focus on a framework that is based on co-evolution, or on the "reconnection of human aspirations and activities with the evolution of natural systems" (Mang & Reed, 2012, p. 26). This framework needs to be 'integral', and part of a holistic model that includes the evolutionary patterns, processes and ecological systems of the natural world. We therefore need to move to the 'higher order of regeneration' as part of nature's evolution, and design place that includes the principles of biophilia (Roös, 2021). A deep connection to place results in providing spaces that include biophilia and topophilia, in essence the inclusion of both the innate human-nature affiliation and the deeper attachment to place that is inclusive of ecological systems and processes and patterns.

*Therefore:*

*Create places in the city that connect to the larger ecological system of the region, as well as connecting local habitats with processes, systems and patterns of place at different levels of scale. Enhance geological and prominent landscape feature connections that contribute to the local biodiversity and habitat. Use biophilic attributes and geometries in design solutions that strengthen the complexity of the systems of local nature.*

## References

Alexander, C. (2001–2005). *The nature of order—An essay on the art of building and the nature of the universe, Book One: The phenomenon of life.* The Center for Environmental Structure.

Alexander, C., Ishikawa, S., Silverstein, M., Jacobson, M., Fiksdahl-King, I., & Angel, S. (1977). *A pattern language: Towns, buildings, construction.* Oxford University Press.

Beatley, T. (2011). *Biophilic cities.* Washington: Island Press.

Browning, W. D., Ryan, C. O., & Clancy, J. O. (2014). *14 Patterns of biophilic design*. Terrapin Bright Green, LLC.

Cole, R. J. (2012). Regenerative design and development: current theory and practice. *Building Research & Information, 40*, 23–48.

Du Plessis, C. (2012). Towards a regenerative paradigm for the built environment. *Building Research & Information, 40*, 5–22.

Du Plessis, C., & Cole, R. J. (2011). Motivating change: Shifting the paradigm. *Building Research & Information, 39*(5), 436–449.

Kellert, S. R. (2018). *Nature by design*. Yale University Press.

Kellert, S. R., Heerwagen, J. H., & Mador, M. L. (Eds.). (2008). *Biophilic design. The theory, science and practice of bringing buildings to life*. Wiley.

Lyle, J. T. (1991). *Design for human ecosystems—Landscape, land use, and natural resources*. Island Press. (Original work published 1985).

Mang, P., & Reed, B. (2012). Designing from place: A regenerative framework and methodology. *Building Research & Information, 40*, 23–38.

McDonough, W., & Braungart, M. (2002). *Cradle to cradle: Remarking the way we make things*. North Point Press.

McHarg, I. (1967). An ecological method for landscape architecture. *Landscape Architecture, 57*, 105–107.

McHarg, I. (1992). *Design with nature*. John Wiley & Sons. (Original work published 1969).

Mehaffy, M. W., Salingaros, N. A., Kryazheva, Y., & Rudd, A. (2020). *A new pattern language for growing regions*. Sustasis Press.

Portoghesi, P. (2000). *Nature and architecture*. Skira.

Roös, P. B. (2021). *Regenerative-adaptive design for sustainable development—A pattern language approach* (Sustainable development goals series). Springer International. https://doi.org/10.1007/978-3-030-53234-5_1

Roös, P. B., & Jones, D. S. (2013). Visions of the surf coast—Changing landscapes under future climate effects. In N. Gurran, P. Phibbs, & S. Thompson (Eds.), *Proceedings of the 10th International Urban Planning and Environment Association Symposium (UPE10—NEXT City: Planning for a New Energy & Climate Future), Sydney, July 24–27, 2010* (pp. 361–372). ICMS Pty Ltd. Retrieved March 9, 2022 from https://www.researchgate.net/publication/261131642_Visions_of_the_surf_coast_-_changing_landscapes_under_future_climate_effects

Salingaros, N. A. (2019). The biophilic index predicts healing effects of the built enviroment. *Journal of Biourbanism, 8*(1), 13–34.

Sampson, S. D. (2012). The topophilia hypothesis: Ecopsychology meets evolutionary psychology. In P. H. Kahn Jr. & P. H. Hasbach (Eds.), *Ecopsychology: Science, tottems and the technological species* (pp. 23–53). MIT Press.

Swaffield, S. (2002). *Theory in landscape architecture: A reader*. University of Pennsylvania Press.

# CHAPTER 7

# Embracing Biophilia: The Pathway to a Regenerative-Adaptive and Resilient Future

**Abstract** Evidence in the literature suggests that the presence of nature in our neighbourhoods and 'green cities' has broader and more pervasive positive impacts on health and wellbeing than thought the case. As evidenced by Beatley (*Biophilic Cities*, Island Press, 2011) in his proposition of *Biophilic Cities*, bringing nature into our cities has many benefits and is good for humans and the environment, as well as supporting a sustainable city agenda. However, moving towards a regenerative-adaptive and resilient future the inclusion of biophilia as a fundamental worldview transitions our current degenerative society to a regenerative society. This chapter investigates the biophilic design pattern framework in relation to sustainability, regeneration and a case for biophilic cities.

**Keywords** Regenerative design • Regenerative-adaptive • Biophilic cities • Biophilic framework • Sustainable design • Resilient

© The Author(s), under exclusive license to Springer Nature
Switzerland AG 2022
P. B. Roös, *A Biophilic Pattern Language for Cities*, Sustainable
Urban Futures, https://doi.org/10.1007/978-3-031-19071-1_7

83

**Fig. 7.1** Urban forest—Botanical Gardens City of Geelong. (Image author: PB Roös, 2022)

## INTRODUCTION

We have identified in the introduction to this book that to apply only the concepts of 'sustainability' is not good enough, and that in our current global crisis of biodiversity loss, major species extinctions and a runaway climate change phenomenon, we need to move 'beyond sustainability' (Fig. 7.1). In the last decade a focus on minimising impacts of human activities on the natural world has resulted in the framework of 'eco-efficiency'. In other words, this means creating more goods and services while using fewer resources (McDonough & Braungart, 2002). This focus on eco-efficiency has translated into the current conceptualisation of the green environmental discourse—emergence of *green design, green planning and sustainable development*—resulting in a dominance of technology solutions and a loss of focus on (social-environmental) systems and wellbeing. Green or eco-efficient design and planning is insufficient because it "misses the real potential that arises out of the human presence on this planet: the possibility of organizing human activities so that they continuously feed and are fed by the living systems within which they occur" (Mang & Reed, 2012, p. 26).

In *Regenerative-Adaptive Design for Sustainable Development* (2021), I investigated various options of how we can move beyond sustainability and the current standard practice of only '100 percent less bad' (after Reed), by integrating the *Integral Sustainable Design Framework* of DeKay (2011) with a *Regenerative-Adaptive Design Model* (Roös, 2021), aiming to move to a 'higher order of regeneration'. The argument is that sustainability is only maintaining the 'status quo', trying to mitigate a more degenerative spiral of planetary resource depletion. The principles of regenerative-adaptive design are being positioned in the regeneration of built and natural environments as a concept where we re-position nature in our settlements, planning and design practice, and resilience planning. This evolution is in part fuelled by the emergence of the Anthropocene discourse, wherein the words of 'living structures', 'generative codes' and 'biophilia' (amongst others) are inspiring new insights into the workings of nature. It is in these concepts and theories that humanity's position and responsibility to the planet Earth are being articulated through the rapidly increasing science of pattern theory linked to a new era in restorative ecology-informed design (Roös & Jones, 2013).

Embedded in this process of regeneration, is where the importance of *biophilia* comes in, as it is necessary to acknowledge that humans are part of nature and need to act as an integral part of the evolutionary, regenerative process of nature itself. Informed by the evolutionary processes of emergence in Alexander's *The Nature of Order* (2001–2005a), this proposition advances a regenerative-adaptive pattern language theory towards a holistic integrated design and planning method. As a result, the *Regenerative-Adaptive Pattern Language* is rooted in the concepts of 'living structures', natural attraction and pattern theory.

Living structures comprises of the geometrical rules of biological forms, a language of patterns; in essence the combination of geometrical properties and elements of landscapes embodied in the complex structures found within all living forms (Salingaros, 2012, 2015, p. 9). *Living structures* support the notion of *biophilia*. Human beings are intrinsically connected and linked to visible and non-visible geometrical forms and patterns in nature; this is evident through the law of natural attraction (Roös, 2021; Roös & Jones, 2017). To include the human-nature affiliation of biophilia in the proposed healing of damaged urban environments, and the design and planning of new cities, the consideration of biophilic design principles and patterns is thus paramount. The *Biophilic Pattern Language for Cities* charts these biophilic patterns and applications to assist with this mammoth task. Although, the end goal is fundamentally important for us if humanity wants to survive—that is to live on the planet in a sustainable future where our cities and places of dwelling are providing outcomes of health and wellbeing, not only for the human, but also for the more-than-human.

## The Nature of Order and Biophilic Patterns for Cities

Indeed, the problems to be solved in a rapidly urbanised world, and to deal with the unprecedented city growth using more and more resources from the Earth, is complex to say the least. It is my proposition that we turn to nature, the system that has sustained itself for more than 3.8 billion years, to find possible answers for the problems we face. It is a complex system; however, it is within this complex system that a *language of patterns* exists, the same pattern language that Christopher Alexander investigated throughout his life. In *The Nature of Order—An Essay on the Art of Building and the Nature of the Universe, Book Two: The Process of Creating Life* (Alexander, 2001–2005b), detailed descriptions and guidance are provided for how to identify living structure in nature, that can be applied to the built environment to create places that feel 'alive', places that have a 'sense of place', and as mentioned in earlier chapters, we need to create places that 'have soul'. Progressing on from Alexander's work on the 15 fundamental properties in the four volume *The Nature of Order* (2001–2005a, b, c, d), and the 253 patterns of *A Pattern Language* (1977), I chart only one key element in this book—that of the *biophilic patterns for cities*—as a part of a much more complex and integrated and integral pattern language for life. Indeed, much more research in this domain of pattern languages is needed, and to document all details of an unending possibility of integral patterns in the dynamics of the human-nature affiliation and regenerative systems, calls for intense future work.

## HIERARCHICAL INTERCONNECTIONS

The *four meta biophilic patterns* listed in this book as part of the *Biophilic Pattern Language for Cities*, are positioned strongly within the one *fundamental pattern* of the *Regenerative-Adaptive Pattern Language*—the pattern *Love for Nature* [6]. The pattern statement notes the following:

> The positive enhancement of connections between humans and their natural environments is fundamental to the health and wellbeing of a global society. Therefore, employ Biophilia to connect people to nature at physical, psychological and consciousness levels. Engage a deep holistic *nature language* that emphasises our love for nature (Roös, 2021, p. 84).

Positioned in a hierarchical framework, at different levels of scale—the four meta biophilic patterns link downwards to various *biophilic attributes* and *elements* as listed in each chapter, and the four meta biophilic patterns link upwards the fundamental pattern *Love for Nature* [6] as listed in the *Regenerative-Adaptive Pattern Language* (2021).

In turn these patterns are connected to other patterns, and it is worthwhile to describe the 15 biophilic design patterns by Downton et al. (2017), adapted from the 14 patterns by Browning et al. (2014). Downton et al. (2017) added an additional pattern, '*Virtual connection to nature*', to the 14 biophilic design patterns to be able to deal with complex constraints in the built environment. These virtual representations of nature could be implemented in design solutions where the provision of actual natural elements such as vegetation is impossible (Downton et al., 2017,

p. 25). When a combination of these patterns is applied to a space, the result is a sense of admiration, and Browning and Ryan (2020) added another biophilic pattern to the list named 'Awe'. This brings the list of biophilic design patterns to a total of 16 patterns. To apply biophilic design in its simplest form, Potteiger and Purinton (1998) and Browning et al. (2014) grouped biophilic design patterns in three categories of reference: *Nature in the Space*; *Natural Analogues*; and *Nature of the Space*. The updated list of 16 patterns is listed in Table 7.1, adopted by Roös (2022) in a recent project (see case study in Chap. 8), and amended to be applied to the contexts and spaces of the urban environment.

The reader will notice that throughout this book in each chapter comparisons have been listed of biophilic patterns by other authors, one set of patterns refers to the patterns listed in Table 7.1. It must be noted that I have only selected specific applicable and corresponding patterns for each *meta biophilic pattern*, to avoid confusion and further complexity.

## THE CASE FOR BIOPHILIC CITIES

Evidence in the literature suggests that the presence of nature in our neighbourhoods and 'green cities' has broader and more pervasive positive impacts on health and wellbeing than has been thought the case. Beatley noted in *Biophilic Cities* (2011) that: "In a national study involving more than ten thousand people in the Netherlands, researchers found significant and sizable relationships between green elements in living environments and higher levels of self-reported physical and mental health"; and "A 2007 Danish study demonstrates the importance of access and proximity to parks and nearby greenspaces: These green features were found to be associated with lower stress levels and lower likelihood of obesity" (Beatley, 2011, p. 6).

In a study led by Peter Newman for the Sustainable Built Environment Research Centre (2012) titled: *Can biophilic urbanism deliver strong economic and social benefits in cities? An economic and policy investigation into the increased use of natural elements in urban design*—key findings indicated that the use of nature in urban design such as green roofs, green walls and indoor plantings, green verges, green islands and green corridors, to urban farming and regenerated waterways has been shown to deliver a range of health and wellbeing benefits when applied throughout cities (SBEnrc, 2012, p. 3). Furthermore, the inclusion of biophilic elements in urban greening findings indicated that:

**Table 7.1** The 16 patterns of biophilic design

| Biophilic design pattern | Context and application |
|---|---|
| *Nature in the space* | |
| **1. Visual and direct connection with nature** A view and direct access to elements of nature, living systems and natural processes. | Ensure visual and direct access to nature such as sensory gardens, plants and trees, animals, water, soils, light, views and (fire)[a]. |
| **2. Non-visual connection with nature** Auditory, haptic, olfactory or gustatory stimuli that engender a deliberate and positive reference to nature, living systems or natural processes. | Enhance opportunities for sensory connections (audible, smell, texture, temperature) to nature through sounds, materials surface treatments (touching), planting and elements that enhance the senses. Include sensory gardens, flowers, season variety. Materiality can include textures from geological materials, smooth to rough (e.g., exposed aggregate). |
| **3. Non-rhythmic sensory stimuli** Stochastic and ephemeral connections with nature that may be analysed statistically but may not be predicted precisely. | Instil patterns of nature's movements and seasonality throughout the outdoor urban scape, aim to connect transition spaces from outdoor to indoor built environments, using formations, forms and shapes, change of time and aging, including artistic representations or installations. |
| **4. Thermal, airflow and light variability** Celebrate natural environments with changes in air temperature, relative humidity, surface temperatures and varying intensities of light and shadow that change over time. | Provide sequential changes in thermal, airflow and light variability through urban scapes, landscapes and transitional spaces from outdoor to internal built environments with shapes and forms, vegetation, materials, texture, colours and natural geometries. (e.g., treatment of surfaces that connects the public realm with entrances to shops, houses, etc.) |
| **5. Presence of water** A condition that enhances the experience of a place through the seeing, hearing or touching of water. | Use water as a static, dynamic and or variable design element to achieve multi-sensory experiences throughout the outdoor urban scapes, and landscapes. |
| **6. Dynamic and diffuse light** Leveraging varying intensities of light and shadow that change over time to create conditions that occur in nature. | Use mixtures of dynamic, diffused and changeable lighting arrangements and patterns (including illuminance and colour) to evoke movement, time, seasonality, while maximising solar access, throughout the outdoor urban scape and landscapes, and at transition spaces from the urban outdoor to the indoor built environment areas. |

(*continued*)

**Table 7.1** (continued)

| Biophilic design pattern | Context and application |
|---|---|
| **7. Connection with natural systems**<br>Awareness of natural processes, especially seasonal and temporal changes characteristic of a healthy ecosystem.<br>*Natural analogues* | Use natural systems (weather, hydrology, geology, in partnership with terrestrial, avian and aquatic wildlife and their diurnal and seasonal patterns) as design inspirations and finishes, throughout the outdoor urban scape and landscapes. |
| **8. Biomorphic forms and patterns**<br>Symbolic references to contoured, patterned, textured or numerical arrangements that persist in nature. | Ensure biomorphic patterns legibility and interest in walls, facades, ground surface places and furniture detail throughout the outdoor urban scapes, and landscapes. (e.g., the inclusion of artworks, benches, seating areas, resting areas). |
| **9. Material connection with nature**<br>Material and elements from nature that, through minimal processing, reflect the local ecology or geology to create a distinct sense of place. | Include the richness of material colour, warmth, authenticity, and tactility throughout the environments. Use of exposed timber or timber finishes extensively throughout the outdoor urban scape and landscapes, and at the transition spaces from the urban outdoor to the indoor built environment areas. |
| **10. Complexity and order**<br>Rich sensory information that adheres to a spatial hierarchy similar to those encountered in Wild Nature. | Prioritise pattern composition and order enabling stimulation, interest and legibility through vegetation planting variability and artwork that results in spatial hierarchy, throughout the outdoor urban scape and landscapes, and at the transition spaces from the urban outdoor to the indoor built environment areas. |
| *Nature of the space*<br>**11. Prospect**<br>An unimpeded view over a distance for surveillance and planning. | Provide a sense of arrival, prospect, and a clear line of sight for each arrival, exit and transition spaces throughout the outdoor urban scape and landscapes, and at the transitional spaces from the urban outdoor to the indoor built environment areas. |
| **12. Refuge**<br>A place for withdrawal, from environmental conditions or the main flow of activity, in which the individual is protected from behind and overhead. | Provide opportunities for retreat, contemplation, waiting, meeting, refuge, in priority areas throughout the outdoor urban scape, and landscapes. |

(*continued*)

**Table 7.1** (continued)

| Biophilic design pattern | Context and application |
| --- | --- |
| **13. Mystery** The promise of more information achieved through partially obscured views or other sensory devices that entice the individual to travel deeper into the environment. | Provide a sense of journey in pedestrian environments that ensures sightlines, permeability and variability in edges and planes, throughout the outdoor urban scape, and landscapes. |
| **14. Risk/peril** An identifiable threat coupled with a reliable safeguard. | Lessen personal risk in preference to safety but do not let safety considerations override biophilic design opportunities and principal executions throughout the outdoor urban scape and landscape. Treat risk/peril elements with Pattern 12—Refuge elements to provide a sense of safety and retreat. |
| **15. Virtual connection with nature** A view to a simulacrum of natural elements, living systems and natural processes. | Provide virtual connections with nature viewed through mediated means or evoked by simulacrums of nature, living systems and natural processes. Examples include live streaming of remote habitats (such as wetlands and forests) to monitors, projectors, electronic boards, throughout the outdoor urban scape, landscapes, and transitioning areas from outdoor to indoor built environments. |
| **16. Awe** Stimuli including a combination of other biophilic design patterns that defy an existing frame of reference and lead to a change in perception. | Create moments of awe through a combination of relevant biophilic design patterns, including works of art that reflect or represent the innate connection to nature (biophilia). Careful consideration of calming and positive effects to be included, surprises that evoke anxiety must be avoided. Provide moments of awe throughout the outdoor urban scape and landscapes. |

Source: Roös (2022); adapted from Downton et al. (2017); derived from Browning et al. (2014)

ª Fire can be symbolic and include warm light, colour and thermal elements included in materials, art installations, façade finishes.

They reduce the urban heat island effect, lessen heating and cooling loads in buildings, improve air quality, allow urban food production, and improve stormwater management. Furthermore, such elements provide aesthetically pleasing surroundings that have been shown to enhance urban liveability, reduce crime and violence, reduce depression, and encourage greater community connectivity. Biophilic urbanism has also been linked to reduced stress, improved health and well-being, increased cognitive abilities, higher productivity, and enhanced early childhood development. (SBEnrc, 2012, p. 3).

Undoubtably, these examples given above clearly indicate that the inclusion of nature in cities is beneficial to both humans and the environment. However, to enhance mental health (as well as physical wellbeing) with the application of biophilic design, and in this instance using the concepts of 'biophilic cities'—more is needed than the 'greening' of the built environment. As I have argued in previous chapters and as advocated by Salingaros (2019), the healing effects of the built environment must include the two distinct sources of:

1. Proximity and visual contact with plants, animals and other people.
2. A response to artificial creations that follow geometrical rules for the structure of organisms (Salingaros, 2019).

These two distinct mechanisms trigger the 'biophilic effect'. As described in Chap. 4, this biophilic effect is a consequence of our biological responses to the natural environment (Murchie, 1978), and this effect of *biophilic patterns* is a result of many *multi-sensory phenomena*. In this book I have attempted to provide the four *meta biophilic patterns* to assist in the design and planning of our cities to achieve this effect, and if applied, should potentially resulting in health and wellbeing outcomes. The inclusion of these patterns in city making therefore supports the case for biophilic cities.

## Conclusion

Throughout this book I have highlighted and stressed the importance of human-nature connection as imperative to support the health and wellbeing of city dwellers. Evidence is clear that embracing biophilia, and changing the way we design our cities by including the principles of biophilic design is undoubtably beneficial for both humans and other species. Further, based on an ecological approach, the interconnectedness of the biophilic patterns and their relevance to sustainability and regenerative practice, should assist in creating places that can potentially result in supporting sustainable cities, as well as supporting the regenerative ecological systems of places, neighbourhoods, cities, regions and ultimately the Earth. The question is how can these biophilic patterns be applied in practice? In the next chapter a case study—*The Green Spine*—in the City of Greater Geelong, Australia, is reviewed to provide guidance to the potential application of the *meta biophilic patterns* as well as the *biophilic design patterns* as noted in Table 7.1.

## REFERENCES

Alexander, C. (2001–2005a). *The nature of order—An essay on the art of building and the nature of the universe, Book One: The phenomenon of life.* The Center for Environmental Structure.

Alexander, C. (2001–2005b). *The nature of order—An essay on the art of building and the nature of the universe, Book Two: The process of creating life.* The Centre for Environmental Structure.

Alexander, C. (2001–2005c). *The nature of order—An essay on the art of building and the nature of the universe, Book Three: A vision of a living world.* The Centre for Environmental Structure.

Alexander, C. (2001–2005d). *The nature of order—An essay on the art of building and the nature of the universe, Book Four: The luminous ground.* The Centre for Environmental Structure.

Alexander, C., Ishikawa, S., Silverstein, M., Jacobson, M., Fiksdahl-King, I., & Angel, S. (1977). *A pattern language: Towns, buildings, construction.* Oxford University Press.

Beatley, T. (2011). *Biophilic cities.* Island Press.

Browning, W. D., & Ryan, C. O. (2020). *Nature inside—A biophilic design guide.* RIBA Publishing.

Browning, W. D., Ryan, C. O., & Clancy, J. O. (2014). *14 patterns of biophilic design.* New York: Terrapin Bright Green, LLC.

DeKay, M. (2011). *Integral sustainable design: Transformative perspectives.* Earthscan.

Downton, P. F., Jones, D. S., Zeunert, J., & Roös, P. B. (2017). *Creating healthy places: Railway stations, biophilic design and the metro tunnel project.* Melbourne Metro Rail Authority, Deakin University.

Mang, P., & Reed, B. (2012). Designing from place: A regenerative framework and methodology. *Building Research & Information, 40,* 23–38.

McDonough, W., & Braungart, M. (2002). *Cradle to cradle: Remarking the way we make things.* North Point Press.

Murchie, G. (1978). *Seven mysteries of life.* Houghton Mifflin.

Potteiger, M., & Purinton, J. (1998). *Landscape narratives: Design practices for telling stories.* John Wiley & Sons.

Roös, P. B. (2021). *Regenerative-adaptive design for sustainable development—A pattern language approach* (Sustainable development goals series). Springer International. https://doi.org/10.1007/978-3-030-53234-5_1

Roös, P. B., & Jones, D. S. (2013). Visions of the surf coast—Changing landscapes under future climate effects. In N. Gurran, P. Phibbs, & S. Thompson (Eds.), *Proceedings of the 10th International Urban Planning and Environment Association Symposium (UPE10—NEXT City: Planning for a New Energy & Climate Future), Sydney, July 24–27, 2010* (pp. 361–372). ICMS Pty Ltd. Retrieved March 9, 2022 from https://www.researchgate.net/publication/261131642_Visions_of_the_surf_coast_-_changing_landscapes_under_future_climate_effects

Roös, P. B., & Jones, D. S. (2017). Weaving landscape fabrics of ecological cities: Patterns for regenerative-adaptive futures. In: *EcoCity World Summit 2017, July 12–14, 2017* (pp. 1–8). EcoCity Builders. Retrieved from http://www.ecocity2017.com/

Salingaros, N. A. (2012). The structure of pattern languages. *Architectural Research Quarterly, 4*(2000), 149–161.

Salingaros, N. A. (2015). *Biophilia and healing environments.* Terrapin Bright Green LLC and Levellers Press. Retrieved from https://www.terrapinbrightgreen.com/wp-content/uploads/2015/10/Biophilia-Healing-Environments-Salingaros-p.pdf

Salingaros, N. A. (2019). The biophilic index predicts healing effects of the built enviroment. *Journal of Biourbanism, 8*(1), 13–34.

SBEnrc. (2012). *Can biophilic urbanism deliver strong economic and social benefits in cities? An economic and policy investigation into the increased use of natural elements in urban design.* Sustainable Built Environment National Research Centre (SBEnrc), Curtin University and Queensland University of Technology.

# CHAPTER 8

# A Case Study: The Biophilic Corridor

**Abstract** What makes biophilic design different to landscape design, landscape architecture or the 'greening' of urban environments? Adding a few pot plants to interior spaces, providing vegetation in the backyard of homes, adding 'greenery' to balconies or including a living green wall to part of the façade of a building—is not biophilic design. In contrary—the fundamentals of biophilic design are embedded in a complex science with deep complexities of human-nature interactions, inclusive of conscious and unconscious responses as humans to natural elements, processes, shapes and forms. This 'living structure', or 'living architecture' is not always visible to the naked eye. This chapter investigates the application of biophilic design patterns to a case study project—debunking the misconception of biophilic design being a simple process of 'greening the city'.

**Keywords** Biophilic design • Green cities • Living architecture • Living structure • Landscape design • Landscape architecture

© The Author(s), under exclusive license to Springer Nature Switzerland AG 2022
P. B. Roös, *A Biophilic Pattern Language for Cities*, Sustainable Urban Futures, https://doi.org/10.1007/978-3-031-19071-1_8

**Fig. 8.1** The Green Spine—City of Greater Geelong. (Image author: PB Roös, 2022)

## Introduction

I am often asked what makes biophilic design different from landscape design, landscape architecture or the 'greening' of urban environments (Fig. 8.1). This is indeed a valid question, as many see adding a few pot plants in interior spaces, design and planting vegetation in the backyard of homes, adding 'greenery' to balconies or include a living green wall to part of the façade of a building—as biophilic design. To be clear, applying a few elements as mentioned above—*is not applying biophilic design in its true sense*. The fundamentals of biophilic design are embedded in a complex science with deep complexities of human-nature interactions, inclusive of conscious and unconscious responses by us as humans to natural elements, processes, shapes and forms (the living structure) that are not always visible to the naked eye (Roös, 2021; Salingaros, 2015). Further, other elements and forces are at play, and this deep connection includes social and cultural contexts. This biophilic effect has been described in Chap. 3 and clearly shows that there are at least two key distinctive mechanisms that are present when biophilia occurs: (1) An intimate contact with other living beings, and (2) A response to geometries and elements that are created by biological rules (Salingaros, 2019). From first observation of many landscape design or greening projects—it is clear in most instances that no deliberate attempt has been made to consider or include these geometries, or the 'living structure' representing the biological rules in nature. In most of the literature it is also evident that focus is put largely on the inclusion of greenery or vegetation in built environments and considered as core to biophilic design (Beatley, 2011; Browning & Ryan, 2020; Kellert et al., 2008). However, I argue that only when the patterns of biophilic design are deliberately and rigorously applied, inclusive of the principles from a pattern language and the 'living structure' of nature, can you truly claim that the outcomes of biophilia has been achieved.

In this chapter a case study has been reviewed to identify the application of biophilic design patterns in an urban environment, in this instance the application to a largely controversial project—The Green Spine, City of Greater Geelong. The review used the 16 patterns of biophilic design as per Table 8.1, and additionally used extracts from the assessment of the project and a design proposal titled—The Biophilic Corridor—done by PhD student Archie Arvindbhai Patel from the School of Architecture, Deakin University (Fig. 8.2) (2020). Finally, the review considers an alignment of the design proposals with the *four meta biophilic patterns* described in this book.

## THE GREEN SPINE: CITY OF GREATER GEELONG

The idea of creating a 'Green Spine' across the main street in the City of Geelong started at an industry workshop hosted by the School of Architecture at Deakin University in 2012, where more than 60 architects, designers, planners and Geelong leaders took part in the 'Vision 2'

**Table 8.1**   The 16 patterns of biophilic design review of the Geelong Green
Spine Block 2

| Biophilic design pattern | Level achieved 0–3 |
|---|---|
| *Nature in the space* | |
| **1. Visual and direct connection with nature** A view and direct access to elements of nature, living systems and natural processes. | 2—Achieved The street is designed as a 'Botanic Walk' with a continuous canopy of trees—Trees as evidence of presence of nature. Layered landscape with a diverse foliage, flowers, colour and texture can attract a lot of variety of birds, butterflies and insects. |
| **2. Non-visual connection with nature** Auditory, haptic, olfactory or gustatory stimuli that engender a deliberate and positive reference to nature, living systems or natural processes. | 1—Partially Achieved Fragrance from the flowering plants—olfactory experience. Chirping of birds and humming of bees— Auditory experience. |
| **3. Non-rhythmic sensory stimuli** Stochastic and ephemeral connections with nature that may be analysed statistically but may not be predicted precisely. | 1—Partially Achieved Cool breeze, sunlight—Tactile experience. Peripheral distraction—Butterflies or bees or rustling of leaves. |
| **4. Thermal, airflow and light variability** Celebrate natural environments with changes in air temperature, relative humidity, surface temperatures and varying intensities of light and shadow that change over time. | 2—Achieved Microclimate—Trees have been found to alter microclimate. Promotes cool breeze. Cleaner quality of air in comparison with other areas with lesser trees/ more manmade intervention. |
| **5. Presence of water** A condition that enhances the experience of a place through the seeing, hearing or touching of water. | 0—Not Achieved Presence of mist/dew drops during dawn—on the grass and the leaves (although this connection with nature is for a limited time of the day). The spaces lack a permanent water feature installation. |
| **6. Dynamic and diffuse light** Leveraging varying intensities of light and shadow that change over time to create conditions that occur in nature. | 1—Partially Achieved Different seasons contribute to different experiences of seasonal quality of light. Different tree structures filter in different qualities of light throughout the spice— Finer textured trees will give more shady places whereas trees with coarse texture will filter in light creating interesting patterns on the floor. Seeking sun during winters and seeking shade under trees during summers. |

(*continued*)

**Table 8.1** (continued)

| Biophilic design pattern | Level achieved 0–3 |
|---|---|
| **7. Connection with natural systems**<br>Awareness of natural processes, especially seasonal and temporal changes characteristic of a healthy ecosystem. | 1—Partially Achieved<br>Presence of deciduous trees gives a sense of different seasons—Some trees will turn bright yellow to orange to bare skeletons and decomposing of leaves during autumn/fall gives us sense of patina of time. Since the spine is an open environment—rain, sun and cold winds will contribute to different seasonal experiences. |
| *Natural analogues*<br>**8. Biomorphic forms and patterns**<br>Symbolic references to contoured, patterned, textured or numerical arrangements that persist in nature. | 0—Not Achieved<br>Even though rounded shaped wooden seating has been provided, there is a lack of biomorphic forms and patterns, and the inclusion of complex geometry is totally absent. |
| **9. Material connection with nature**<br>Material and elements from nature that, through minimal processing, reflect the local ecology or geology to create a distinct sense of place. | 1—Partially Achieved<br>Use of wood—minimally intervened—kept raw—in the benches. Use of natural material in some parts of the paving. |
| **10. Complexity and order**<br>Rich sensory information that adheres to a spatial hierarchy similar to those encountered in Wild Nature. | 1—Partially Achieved<br>Fractal geometry of the planting palette achieved only a limited representation of complexity and order. |
| *Nature of the space*<br>**11. Prospect**<br>An unimpeded view over a distance for surveillance and planning. | 1—Partially Achieved<br>The continuity of the spine alongside the edge of the street gives a distinct view of long distance and sight, giving a sense of 'openness'. |
| **12. Refuge**<br>A place for withdrawal, from environmental conditions or the main flow of activity, in which the individual is protected from behind and overhead. | 1—Partially Achieved<br>Trees and shrubs provided shelter and a sense of protection overhead. However, deliberate resting places with shelter for withdrawal from the busy street have not been provided. |
| **13. Mystery**<br>The promise of more information achieved through partially obscured views or other sensory devices that entice the individual to travel deeper into the environment. | 1—Partially Achieved<br>The meandering walkway lets you explore in bits and pieces and awakens a sense of curiosity to further investigate the space. |

*(continued)*

**Table 8.1**  (continued)

| Biophilic design pattern | Level achieved 0–3 |
|---|---|
| **14. Risk/peril**<br>An identifiable threat coupled with a reliable safeguard. | 0—Not Achieved<br>This pattern is not evident in the spine. |
| **15. Virtual connection with nature**<br>A view to a simulacrum of natural elements, living systems and natural processes. | 0—Not Achieved<br>This pattern is not evident in the spine. |
| **16. Awe**<br>Stimuli including a combination of other biophilic design patterns that defy an existing frame of reference and lead to a change in perception. | 1—Partially Achieved<br>Even though the spine provides new vegetation and walking paths along Malop street, the findings are that biophilic patterns were partially applied, and that the presence of the biophilic patterns was limited. Also, noticeable is that the existing trees are sparse and scattered, resulting in that the 'human-made' world seems to be overpowering the natural elements. |

Source: Roös (2022)

**Fig. 8.2**   The biophilic corridor. (Image author: Patel A, 2020)

planning process to reshape the future of Geelong's city centre. The project leader at the time, and Head of School of Architecture at Deakin— Professor Hisham Elkadi, coined the phrase Green Spine' to describe a linear park that he imagined running through the CBD from the train station to the botanical gardens (Geelong Advertiser, 2020a). Ten years later the idea of the implementation of the 'green spine' is in its second phase; this is to change a street dominantly for vehicle use into a mixed walkable street with added greenery and vegetation. The project was divided into six blocks of project work, currently (as of September 2022) with Block 2 completed, Block 1 partially completed and Block 3 at concept design stage with community consultation (RCG, 2022).

However, the project experienced various issues and resistance and in June 2019 Cr Eddy Kontelj called for changes to the Green Spine, the removal of bike lanes and the return of vehicle lanes and on-street parking. Under pressure, in February 2020 the Geelong City Council decided to reintroduce turning lanes, increase parking spaces and demolish one of two separated bike lanes that were part of the $8 m joint council-government Green Spine project, because councillors claimed it was causing traffic congestion and disability access problems. Things turned bad, and the Victorian State Government officially took control of the project, reinstating the vision and programme to complete the Green Spine (Geelong Advertiser, 2020b). The Green Spine project was included as part of the $500 million Geelong City Deal by the State Government, a collaborative plan to transform Geelong and the Great Ocean Road, by the Australian and Victorian Governments and the City of Greater Geelong through the Revitalising Central Geelong Action Plan (RCG, 2022).

Notably, the project fact sheet for Block 3—Yarra Street to Bellarine Street, mentions that biophilic design principles are key to the project, and states that "A key feature in the designs is the use of 'biophilic' design principles, seeking to connect buildings and people more closely with nature. Biophilic city design has been proven to deliver a range of environmental, social and economic benefits to the community" (RCG, 2022).

The review of this project is therefore timely, considering that it is publicly noted to include biophilic design principles in the next phases of the project.

# A CRITICAL REVIEW

In the first instance when looking at the completed section of the Green Spine (Block 2) and walking along the spine between Moorabool and Yarra streets, the impression is that the inclusion of nature in the city is quite successful. The Victorian Government claimed that the Green Spine is "...a vibrant street and linear park, with an 8-metre-wide botanical walk, featuring stunning indigenous and exotic plants and a series of interconnected gathering spaces beneath a canopy of large trees. It provides more spaces for people to meet and socialise. The separated one-way bike lanes and meandering walkways provide for safe active movement and transport, supporting the health and well-being of the community" (RCG, 2022).

However, the question is how many of the 16 biophilic design patterns have been considered and successfully implemented, and what opportunities exist to improve the inclusion of biophilic design in future projects? To do a critical review is a straightforward and simple exercise. Since 2018 students participated in an assignment that reviewed and assessed the Green Spine considering biophilic design principles—part of an elective unit for architecture and landscape architecture—*SRP761 Ecological Cities and [Biophilic] Futures*—that I developed and lectured in. In Table 8.1 findings of this work have been consolidated, interpreted and rated by students providing one overall result between 0 and 3 where: 0—Not Achieved, 1—Partially Achieved, 2—Achieved and 3—Overachieved/ Excelled. Comments made on their observation of the patterns have been extracted from one student's work—Archie Arvindbhai Patel—in combination with other comments that reflected a majority view of the analysis by all the students (Patel, 2020).

It is worthwhile noting that this stage of the project (Block 2) has not been informed by biophilic design, or any design or implementation deliberately guided by biophilic design principles. It is only in the next design stages of the project that the Victorian Government noted a consideration for biophilic design. The question remains how serious this statement is, and if the actual 'design science' of biophilic design will be applied and included in the project. As noted in the Peter Newman (2012) study, various barriers still exist in the application and adoption of biophilic design in city planning (SBEnrc, 2012).

The critical review indicated that elements of landscape design and the 'greening' of a city scape are evident by default and correlate with some of the biophilic design patterns such as 'Visual Connection with Nature [1]', and 'Thermal, Airflow and Light Variability [4]'. This is to be expected as vegetation and outdoor areas by default have many of these two biophilic attributes present. However, when we look at the more complex patterns to be considered, it indicates very quickly that current design practice lacks the breadth and depth, as well as adequate knowledge of biophilic design science. For example, there is a noticeable lack of intervention in built structures indicating the 'living structure', created through the representation of ornament and geometrical shapes and forms—these are avoided or considered too difficult to implement. As a result, and from this critical review, it is evident that the project does not include sufficient biophilic design principles to claim a status of designed by 'biophilic design'. However, as indicated in Table 8.1, the outcomes of the project improved the streetscape significantly, and there will be positive results for the users with the presence of a more prominent green space in the city. So how can this Green Spine be elevated to the next level where all the biophilic design patterns can be included? In the next section design solutions are proposed for 'The Biophilic Corridor' by Patel (2020).

## A Future Vision: The Biophilic Corridor

In this proposal by Patel (2020) the author noted that the idea was to improve, implement and improvise on a design proposal to undo the flaws in the existing Green Spine project, to ensure a more 'biophilic' outcome. The proposal intended to identify the partially evident biophilic patterns and introduce the missing ones. In addition, the design proposal attempted to homogenise the occurrence of patterns that are scattered and limited to specific areas, in order to be able to promote a more holistic experience, and a deeper connection to nature (Patel, 2020, p.14).

To improve the 'innate experience of nature in the city' three components were proposed to complement the existing Green Spine, a 'layered plantation', 'elevated walkways' and 'nesting pods'. After detailed analysis of the existing conditions, the design proposal recommended removing vehicle use in the main street and rerouting vehicles, so as to be able to allow for a walking and biodiversity corridor from Johnstone Park to Eastern Park, next to the botanical gardens. The next step included a new 'plantation' scheme (Fig. 8.3), that included *Component One—A Layered Plantation*. The rationale behind the layered plantation (a vertical layered scheme) was inspired by the elements of a rainforest. Each of the layers are unique, and in each layer different species of insects and birds are present.

**Fig. 8.3** Component one—new plantation scheme biophilic corridor. (Image author: Patel A, 2020)

The outcome is that each layer renders different qualities for human nature interactions. As part of the layering process and design, diverse local species have been identified to help to recreate the rainforest typology.

The second component of the biophilic intervention considered the experience of an 'elevated nature'. This can be compared to the discussions by Beatley (2011) about the *nature above* and the *nature all around* (Beatley, 2011, pp. 18–31). *Component Two—Elevated Walkways* proposed elevated walkways that complement the layered plantation (Fig. 8.4). With the walkways at varying heights throughout the tree canopies and other levels, a range of experiences can be explored. In some areas, the walkways broaden out to provide gathering spaces where people can sit, relax and reflect in the most dense part of the city still able to experience a natural environment elevated above other activities on the ground (Patel, 2020, p.20).

The third component of the biophilic intervention considered opportunities along the corridor to pause and reflect, resting places, as well as to provide places of refuge. *Component Three—Nesting Pods* proposes an extension of current resting areas, but more so, a continuation of the nesting of local bird species. This intervention can be compared to the requirements stated by Salingaros (2019), that there is a need to include nature's geometry in built forms, supporting the requirement of *complex geometry* to allow for a *biophilic effect* (Salingaros, 2019). The shape of these pods represent weaver bird nest-like structures, proportioned to allow for human use within the elevated walkway (Fig. 8.5). The nesting pods allow a person to seek shelter or to engage in restorative activities such as reading, meditation and even play (Patel, 2020, p. 22).

## Conclusion

How does the assessment outcome of the *Green Spine* project and this proposal of the *Biophilic Corridor* align with the *four meta biophilic patterns* chartered previously in this book? In the first instance we can see that various biophilic attributes, resulting from the 16 biophilic patterns (Table 8.1), indicate that the *four meta biophilic patterns* are supported, but only partially. It is obvious that an in-depth study of the principles of the biophilic patterns is needed, and needless to say—education around biophilia and biophilic design for design practitioners has become a matter of urgency. We need architects, landscape architects, planners, urban designers and city managers to take an urgent, bold step forward to move

**Fig. 8.4** Component two—elevated walkways biophilic corridor. (Image author: Patel A, 2020)

beyond the current standard approach to city making—embracing innovative design solutions. The proposal by Patel (2020) is an innovative design approach as an example for design practitioners, and a bold presentation of the intervention to include biophilic design principles to enhance our

**Fig. 8.5** Component three—nesting pods biophilic corridor. (Image author: Patel A, 2020)

public areas for better health and wellbeing in the urban scape of the City of Geelong. The biophilic assessment of the Geelong Green Spine by the students of *SRP761 Ecological Cities and Biophilic Futures* indicated that there is huge opportunity for improvement, and ultimately the conclusion was drawn that the understanding of *biophilia* and the *science of biophilic design* in our built environment is in its infancy.

It is evident that many attempts are made to include nature in city spaces, and in some instances we can see that some biophilic design patterns are addressed by some design interventions. We can also state that there are benefits from these interventions, and undoubtably there are improvements in our cityscapes when greenery is included. These benefits include the reduction of the urban heat island effect, less heating and cooling loads, improved air quality and providing spaces for community recreation, just to name a few. It is also possible to claim that greening of our cities supports a sustainable future and is key in addressing the UN Sustainable Development Goal (SDG)11—Sustainable Cities and Communities. However, will these standard landscape designs and the greening of cities provide all the necessary benefits we need for the improvement of health and wellbeing of our city dwellers? From the evidence in the literature, I'll say no...

There is an urgent need to embrace *A Pattern Language* (Alexander et al., 1977) approach when designing and planning our cities, and a call to apply the *Biophilic Pattern Language for Cities* as charted throughout the chapters of this book.

## REFERENCES

Alexander, C., Ishikawa, S., Silverstein, M., Jacobson, M., Fiksdahl-King, I., & Angel, S. (1977). *A pattern language: Towns, buildings, construction*. Oxford University Press.

Beatley, T. (2011). *Biophilic cities*. Island Press.

Browning, W. D., & Ryan, C. O. (2020). *Nature inside—A biophilic design guide*. RIBA Publishing.

Geelong Advertiser. (2020a). *Timeline of Geelong's Green Spine Project*. Geelong Advertiser. Retrieved September 1, 2022 from https://www.geelongadvertiser.com.au/news/geelong/timeline-of-geelongs-green-spine-project/news-story/238c3c7a850b37916266c6288d2c72dc

Geelong Advertiser. (2020b). *Malop St back under State rule afer Geelong council's Green Spine conttroversy*. Geelong Advertiser. Retrieved September 1, 2022 from https://www.geelongadvertiser.com.au/news/geelong/malop-st-back-under-state-rule-after-geelong-councils-green-spine-controversy/news-story/dca532152ef6ff20519efc7a01213506

Kellert, S. R., Heerwagen, J. H., & Mador, M. L. (Eds.). (2008). *Biophilic design. The theory, science and practice of bringing buildings to life*. Wiley.

Patel, A. (2020). *The Biophilic Corridor*. Unpublished SRP761 Ecological Cities and Futures Assignment, Deakin University, Geelong.

RCG. (2022). *Green Spine. Revitalising Central Geelong Projects, Revitalising Central Geelong (RCG)*. Victoria State Government. Retrieved September 1, 2022 from https://www.revitalisingcentralgeelong.vic.gov.au/projects/underway-projects/green-spine-future-blocks

Roös, P. B. (2021). *Regenerative-adaptive design for sustainable development—A pattern language approach* (Sustainable development goals series). Springer International. https://doi.org/10.1007/978-3-030-53234-5_1

Salingaros, N. A. (2015). *Biophilia and healing environments.* Terrapin Bright Green LLC and Levellers Press. Retrieved from https://www.terrapinbright-green.com/wp-content/uploads/2015/10/Biophilia-Healing-Environments-Salingaros-p.pdf

Salingaros, N. A. (2019). The biophilic index predicts healing effects of the built enviroment. *Journal of Biourbanism, 8*(1), 13–34.

SBEnrc. (2012). *Can biophilic urbanism deliver strong economic and social benefits in cities? An economic and policy investigation into the increased use of natural elements in urban design.* Sustainable Built Environment National Research Centre (SBEnrc), Curtin University and Queensland University of Technology.

# Index[1]

---

[1] Note: Page numbers followed by 'n' refer to notes.